Ming Liu and Z.Jun Lin

Corporate Governance, Auditor Choice and Auditor Switch

Ming Liu and Z.Jun Lin

Corporate Governance, Auditor Choice and Auditor Switch

Evidence from China

VDM Verlag Dr. Müller

Impressum/Imprint (nur für Deutschland/ only for Germany)
Bibliografische Information der Deutschen Nationalbibliothek: Die Deutsche Nationalbibliothek
verzeichnet diese Publikation in der Deutschen Nationalbibliografie; detaillierte bibliografische
Daten sind im Internet über http://dnb.d-nb.de abrufbar.

Coverbild: www.purestockx.com

Verlag: VDM Verlag Dr. Müller Aktiengesellschaft & Co. KG
Dudweiler Landstr. 99, 66123 Saarbrücken, Deutschland
Telefon +49 681 9100-698, Telefax +49 681 9100-988, Email: info@vdm-verlag.de
Zugl.: Hong Kong, Hong Kong Baptist University, Diss., 2007

Herstellung in Deutschland:
Schaltungsdienst Lange o.H.G., Berlin
Books on Demand GmbH, Norderstedt
Reha GmbH, Saarbrücken
Amazon Distribution GmbH, Leipzig
ISBN: 978-3-639-10868-2

Imprint (only for USA, GB)
Bibliographic information published by the Deutsche Nationalbibliothek: The Deutsche
Nationalbibliothek lists this publication in the Deutsche Nationalbibliografie; detailed
bibliographic data are available in the Internet at http://dnb.d-nb.de.

Cover image: www.purestockx.com

Publisher:
VDM Verlag Dr. Müller Aktiengesellschaft & Co. KG
Dudweiler Landstr. 99, 66123 Saarbrücken, Germany
Phone +49 681 9100-698, Fax +49 681 9100-988, Email: info@vdm-publishing.com

Printed in the U.S.A.
Printed in the U.K. by (see last page)
ISBN: 978-3-639-10868-2

Table of Contents

Chapter One
Introduction

1.1 Research Motivations

The quality of corporate financial disclosure has become an important policy issue following the Enron and WorldCom scandals. Enhancing disclosure quality increases transparency which facilitates public investors to better monitor firms. Fan and Wong (2002) argue that low transparency is associated with high agency costs and low level of corporate governance. Audits conducted by independent professionals, however, can serve a very important role in improving financial disclosure and information credibility (Balsam et al. 2003; Ferguson et al. 2004). Nevertheless, the quality of auditors will also affect the utility of financial audits. The purpose of this study is to investigate the association between firms' internal corporate governance mechanism and their auditor choice and auditor switch decisions and how investors respond to firms' auditor choice and auditor switch decisions in the Chinese context.

Auditor choice, or client-auditor alignment, can be viewed as the minimum cost match between client needs (the demand side) and auditor services (the supply side) in a certain auditing environment. Studies on auditor choice and auditor switch to date have been conducted predominantly in the US (Hudaib and Cooke 2005; Lee et al. 2004; Pittman and Fortin 2004; Copley and Douthett 2002; Geiger et al. 1998; Copley et al. 1995; Krishnan 1994; Johnson and Lys 1990; DeAngelo 1982; Chow and Rice 1982), with occasional studies in countries such as Australia (Craswell 1988), New Zealand (Firth 1999; Firth and Smith 1992) and the UK (Beattie and Fearnley 1995), where the auditing environments are fairly similar. One reason why audit markets in these countries have been studied extensively relates to the developed capital markets in these countries. However, there are few empirical studies that examine auditor choice and auditor switch in the emerging economies. Emerging economies have less developed equity markets and very different auditing environment than the developed ones (Woodward 1997). This study extends the auditor choice and auditor switch literature further from the developed capital markets to the less developed Chinese market.

Shortly after the founding of the People's Republic of China in 1949, the auditing profession in China diminished completely. Independent audits were virtually nonexistent under the planned economy before the 1980s, when the State both owned and ran enterprises. The re-emergence of independent

1

auditing was the result of mushrooming Sino-foreign joint ventures, brought about by China's open-door policy adopted in the early 1980s. Due to non-state-owned interests in the joint ventures, demand emerged for the verification of capital contributions and audits of annual financial statements and income tax returns by registered non-government-employed Chinese certified public accountants (Xiao et al 2000). The progress of full-scale economic reforms, with the separation of ownership and management of enterprises, leads to agency problems in business firms. Independent audits are thus called for to alleviate the agency problems in China. The shift of ownership rights from the state to private and institutional investors in pace with increasing diversification in the economy has further promoted the monitoring role of independent audits.

The Chinese Institute of Certified Public Accountants (CICPA) was established in the early 1980s. At the turn of 1990s, the rapid development of shareholding companies (stock companies) in China led to a sharp increase in the demand for external audits. The establishment of the Shanghai and Shenzhen stock exchanges in 1990 and 1991 respectively and the promulgation of new accounting and auditing standards have played an important role in this process. The China Securities Regulatory Commission (CSRC) requires that all listed firms have their annual reports audited by certified public accountants (CPAs). The monitoring of both public and private enterprises by independent auditors has been employed by the government as an important mechanism in transforming the Chinese economy from the one directed by the "visible hand" of centralized planning to the one guided by the "invisible hand" of market forces.

As the largest and fastest growing emerging market, China has become more and more important to investors throughout the world. However, the Chinese market is far from perfect. The stock market in China is characterized by a less rigorous enforcement of financial reporting requirements compared with the developed economies. In China, the stock market has quite different features, for example, there exists very little pre-disclosure information such as earnings forecasts or preliminary earnings announcements[1] compared with that in the US. In China, the credibility of financial statements was perceived low, and auditors were perceived to be lack of independence and professionalism. A survey published in the China Securities Daily (May 16, 1995) shows that 60 percent of investors had doubts in the audited financial statements. Furthermore, a lot of small investors have very limited knowledge

---

[1] Since 1999, the Chinese Securities Regulatory Commission (CSRC) has required the listed firms to make a preliminary announcement if it is expecting an annual loss. Normally the announcement only states that an annual loss is expected and provides no information about its magnitude.

and experiences in using financial information. Because of the uniqueness of the capital market and the auditing profession in China, it is of a great interest to study auditor choice and auditor switch issues in such an emerging market.

With the separation of shareholders and managers come the opportunistic management behaviors and severe agency costs. Investors, especially the small investors of publicly held firms, typically neither manage nor control the operations of the firms. To reduce risk exposure, they tend to diversify their holdings across a variety of firms as a hedge against potential financial losses. As a result, small investors have little interest in conducting, or even closely monitoring, the day-to-day activities in the firms in which he or she has a financial interest (Fama 1980). Thus the principal-agent relationship arises between the owners/shareholders and the management. The principal-agent relationship, depicted in agency theory, is important in understanding how the audit has developed. The principals entrust resources to the agents and delegate some decision-making authority to the agents. In so doing, the principals place trust in their agents to act in the principals' best interests. However, as a result of information asymmetries between the principals and the agents and of different motivations, the agents may break the contractual trust at the expense of the principals. Hence, the principals may lack trust in their agents and need to put in place monitoring mechanisms, such as the audit, to reinforce this trust. This created a market for independent auditors, hired by firms to provide a check on the managers' performance with the owners' resources (Imhoff 2003).

Independent auditors, by performing their audits in accordance with the Generally Accepted Auditing Standards (GAAS), would attest the fairness of management's financial reports to the stakeholders, and would discover any deviations from the Generally Accepted Accounting Principles (GAAP). Firms can thus employ reputable auditors to assure the shareholders and the potential investors of the credibility of accounting information and hence mitigate the agency problems underlying the contractual relations between firm management and the owners/shareholders. In attesting the credibility of accounting information, independent auditors thus serve an external monitoring role on behalf of the owners/shareholders (e.g., Fan and Wong 2005; Ashbaugh & Warfield 2003). Audits can reduce agency costs by assuring the quality of financial statements, thereby allowing more precise and efficient contracts between the principals and the agents to be based on the financial statements (Watts and Zimmerman 1986; Bedard and Johnstone 2004).

According to OECD's definition, corporate governance is the system by which business corporations are directed and controlled. The corporate governance structure specifies the distribution of rights and responsibilities among different participants in the corporation, such as the board, managers, shareholders and other stakeholders, and spells out the rules and procedures for making decisions on corporate affairs. In so doing, it also provides the structure through which the company objectives are set, and the means of attaining those objectives and monitoring performance. Corporate governance mechanisms and controls are designed to reduce the inefficiencies that arise from moral hazard and adverse selection, including internal corporate governance controls and external corporate governance controls.

Internal corporate governance controls monitor operating activities and then take corrective actions to accomplish the organizational goals. Examples of internal corporate governance controls are of monitoring by and for the BoD, balancing the right or interest of the controlling and minority shareholders, and reviewing performance-based remuneration for managers. The effectiveness of internal corporate governance controls is thus determined by the organizational arrangement for the internal monitoring mechanisms, including ownership concentration, supervision over the BoD and management, duality of positions for business executives and monitoring authority, and so on. External corporate governance controls encompass the controls exercised by external stakeholders over the organization. Examples include debt covenants, external audits, government regulation, media pressure, takeovers, competition, and managerial labor market. Auditors provide an independent check on the work of the management and on the quality of the financial information provided by the management, and therefore serve a fundamental role in promoting confidence and reinforcing trust in corporate financial reporting. By performing the attestation function, auditing is a significant part of a firm's monitoring system. In this way, as a necessary mechanism for the function of contractual relations, audits are an essential component of the corporate governance mosaic.

Firms with high agency costs are more inclined to choose a high-quality auditor to improve their corporate governance (Fan and Wong 2005; Hay and Davis 2004). One major benefit from improved corporate governance is firms can raise capital at lower costs (Bloomfield 2004). The listed firms in China are featured with high ownership concentration and therefore lower level of transparency (i.e., opaqueness), which leads to information asymmetry between the controlling shareholders and the other shareholders. Theoretically, the level of opaqueness is associated with the effectiveness of corporate governance mechanism, both internal and external. The more effective (stronger) corporate governance mechanism a firm has, the less opaque it is; vice versa. In this regard, with their monopolist status, the

4

controlling shareholders are able to derive certain opaqueness gains --- private benefits expropriated from other shareholders because of information opaqueness. Therefore, there are incentives for firms with high ownership concentration to opt for or against high-quality audits. On the one hand, firms hiring high-quality auditors could signal good corporate governance to outside investors, so they can raise capital through the equity or debt market at lower costs. On the other hand, the controlling owners may lose their opaqueness gains if firms are monitored by a high-quality auditor, e.g., it will be more difficult for the controlling shareholders to benefit from earnings management and tunneling behaviors (the transfer of resources out of a firm to its controlling shareholder).

In this study we choose a specific period of 2001-2004, when China experienced a continuous span of bear market. During 2001-2004, the listed firms were unlikely to issue equity to the public as both stock prices and market confidence were low. CSRC even suspended the listed firms from issuing new equity in 2002. Since issuing new equity is unlikely, benefits from lowering capital raising costs will correspondingly be trivial. Then, the major concern in choosing and switching auditors derives from the opaqueness gains. Since firms with weaker corporate governance mechanism are likely to have more opaqueness gains (Fan and Wong 2002), the four-year bear market in China provides a good opportunity to study the association between firms' internal corporate governance mechanism and their auditor choice and auditor switch decisions.

Considerable empirical evidence supports a contemporaneous correlation between accounting earnings and stock price changes (e.g., Holthausen & Verrecchia 1988; Abarbanell and Lehavy 2003). However, earnings explain only a small portion of the variation in market returns at the earnings announcement date. This has led to a search for models to incorporate "non-earnings" information. Auditor choice and auditor switch are found to be important information in explaining the movement of market returns in the developed audit markets (Pittman and Fortin 2004; Citron and Manalis 2001; Chan et al. 2006; Teoh 1992). In this study, we extend the research issue to China, where the market is perceived to be of low efficiency and investors are relatively inexperienced.

Several factors motivated this study. First, the audit market in China presents an interesting arena for the study of auditor choice and auditor switch. Different from the developed economies, the Chinese accounting and auditing professions are not only regulated but also administered by government agencies (Chen et al. 2000). The government exercises control by setting the professional standards and by directly monitoring the operation of auditing firms. It is therefore of great interest to examine

5

whether independent audits are value relevant in a market where stringent government control often prevails over market mechanisms. Second, CICPA has recently begun to rank auditors in China in order to improve the transparency and quality of the Chinese auditing market, which presents an opportunity for identifying high-quality auditors in the Chinese context. An empirical study of the audit quality and firms' auditor choice and switch is therefore useful. Third, for controlling owners of the listed firms, there is always a tradeoff between hiring a high-quality auditor to improve corporate governance and hiring a low-quality auditor to sustain the opaqueness gains from relatively weak corporate governance mechanism, e.g, benefits through tunneling behaviors. Usually it is difficult to disentangle these two motivations. However, the bear market in China during 2001-2004 provides a good opportunity to study the association between internal corporate governance mechanism and firms' decisions on auditor choice and auditor switch. Fourth, market implications of auditor choice and auditor switch in China are perceived different because the incentives for earnings management in China are of distinctive nature. Unlike in the US, the incentives for earnings management in China are dominantly in the direction of increasing income. In this regard, we expect some different findings on investors' responses toward firms' decisions on auditor choice and auditor switch in the Chinese context.

1.2 Major Findings

The empirical results show that firms with larger controlling owners, with smaller size of the supervisory boards (SB), or in which the CEO and the chairman of the board of directors (BoD) are held by the same person are less likely to hire a Top 10 auditor in China. This suggests that when benefits from lowering capital raising costs are trivial, firms with weak corporate governance mechanism would decline to choose a high-quality auditor so as to protect their opaqueness gains.

Results of auditor switch tests show that firms with larger controlling owners, or in which the CEO and the BoD chairman are held by the same person are more likely to switch to a smaller auditor rather than to a larger auditor. However, the SB size has an insignificant effect. This may suggest that the monitoring role of SB is not consistently effective in terms of corporate governance functioning. The reason might be that SB members are mostly from inside the firm and SB members do not have enough expertise in corporate governance. The finding suggests that firms with weak corporate governance mechanism tend to switch to a more pliable auditor to sustain the opaqueness gains.

6

The market implications model shows that for firms with positive abnormal earnings, auditor quality and switching to a larger auditor have a positive impact on earnings response coefficients (ERCs)[2] and switching to a smaller auditor has a negative impact on ERCs. Vice versa, for firms with negative abnormal earnings, both auditor quality and switching to a larger auditor have a negative impact on ERCs and switching to a smaller auditor has a positive impact on ERCs. The results suggest that Top 10 auditors may be perceived more effective in curbing income-increasing earnings management, leading to higher ERCs for clients with good news and lower ERCs for clients with bad news. Firms' switching to a larger auditor may signal conservatism on financial reporting, investors therefore appreciate their stock prices more for positive abnormal earnings and depreciate their stock prices less for negative abnormal earnings. Vice versa, switching to a smaller auditor may signal aggressiveness on financial reporting and hence results in market responses in the opposite direction.

The evidence, in general, suggests that audit-related information in China is valued by the market and large auditors have been able to product-differentiate them in the Chinese market. The issue of audit quality and confidence in corporate disclosure is identified as an important factor for the Chinese market, in which investor confidence has to be bolstered in order to mobilize resources for the private business sectors. Our findings may also generate implications for international investors. As CSRC recently granted licenses to Qualified Foreign Institutional Investors (QFII) such as Morgan Stanley, Goldman Sachs, and Citibank to participate directly in China's A-share market, the findings suggest that international investors need to be aware of the structural arrangement of corporate governance of the listed firms and the effectiveness of audit services in China.

1.3 Contributions

In firms with high ownership concentration, controlling owners have a tradeoff between improving corporate governance and reaping opaqueness gains from relatively weak corporate governance mechanism. While prior research does not disentangle the two motivations (improving corporate governance to lower capital raising costs and reaping opaqueness gains), our study, by carefully choosing a time period to minimize incentives for lowering capital raising costs, intends to specifically

---

[2] Earnings Response Coefficients (ERCs) measure the relation between accounting earnings and stock returns. ERCs are commonly estimated as the slope coefficient in a regression of the abnormal stock returns on a measure of earnings surprise.

examine the association between firms' auditor choice and auditor switch decisions and their internal corporate governance mechanism.

Prior auditor switch studies have examined the relation between firms' auditor switch and their firm-specific characters. However, different types of switch may imply for different incentives and have different market implications. In this study, we identify two major types of auditor switch, namely switching to a larger auditor and switching to a smaller auditor. As the financial statements audited by large auditors are perceived to be of high quality and credibility (e.g., Chan and Wong 2002), the two switch types may have opposite signaling effects. The empirical results turn to support our hypotheses regarding the two different types of auditor switch. For firms that have switched auditors, those with relatively weak corporate governance mechanism are inclined to switch to a successor auditor with lower quality.

Studies on market implications of audit quality (e.g., Krishnan 2003; Kothari 2001; Holthausen and Verrecchia 1998) document higher ERCs for firms audited by big auditors. We expect that the impact of audit quality on ERCs will be different depending on the direction of abnormal earnings. For positive abnormal earnings, higher ERCs imply a greater appreciation of stock prices; while for negative abnormal earnings, higher ERCs imply a greater penalty of stock prices. As the motivation for earnings management in China (and also in many other emerging markets) is characterized by a nearly sole direction of increasing income, auditors' role in curbing earnings management is perceived dominantly in the direction of reducing management-reported income. Separately testing the impact of auditor quality and auditor switch on ERCs for firms with positive and negative abnormal earnings provides more convincing results. Findings demonstrate that firms audited by Top 10 auditors have higher (lower) ERCs if they report positive (negative) unexpected earnings; that firms switching to a larger auditor have higher (lower) ERCs if they report positive (negative) unexpected earnings; and that firms switching to a smaller auditor have lower (higher) ERCs if they report positive (negative) unexpected earnings.

1.4 Thesis Structure

The thesis is organized as follows: in Chapter 2 we use logit regression to test the relation between the internal corporate governance mechanism and auditor choice decisions; in Chapter 3 we examine the

8

association between the internal corporate governance mechanism and the types of auditor switch; in Chapter 4 we examine market responses to firms' decisions on auditor choice and auditor switch; and in Chapter 5 we conclude the study, discuss the limitations, and suggest future research.

Chapter Two

The Impact of Internal Corporate Governance on Auditor Choice

2.1 Literature Review

2.1.1 Audit Utility and Audit Quality

Prior audit research uses the agency costs to explain heterogeneous demand for external audit services (DeAngelo 1981; Dye 1993; Chaney et al. 2004). Due to the varied nature of agency costs, there are heterogeneous demands for independent audits to serve as a monitoring function, characterized by different levels of audit quality. Audit quality has been defined in different ways. These definitions embrace, to varied extent, the dimensions of competence and independence of auditors. A high-quality auditing firm should have independence (relationship based), enough expertise (technique based), and good integrity (honesty and forthrightness). In a broad sense, auditors' independence includes expertise and integrity (Pittman and Fortin 2004; Schauer 2002).

Most empirical research defines the audit quality in relation to the audit risk that an auditor may fail to modify the opinion on the financial statements that are materially misstated. To modify his/her opinion on the financial statements, an auditor needs to have the expertise to find out the misstatement and be willing to report it. DeAngelo (1981) has defined auditor independence as the joint probability that auditors will find out and report misstatements in the financial statements. Using ERCs from returns-earnings regression as a proxy for investors' perceptions of earnings quality, Ghosh and Moon (2005) found that investors and information intermediaries perceived auditor tenure as an indicator of audit quality. This finding may suggest that longer tenure leads auditors to better understand their clients' businesses and to have more expertise to find out misstatements. Lennox (2005) found that firms with top management affiliated with their auditors are significantly more likely than unaffiliated firms to receive clean audit opinions. This may imply that affiliation with clients makes auditors less willing to report misstatements. Although diverse to certain extent, most existing definitions of audit quality reflect some aspects of the DeAngelo's definition.

DeAngelo (1981) argues that audit firm quality is positively associated with firm size or the firm's market share. She argues that audit firms are faced with countervailing incentives regarding independence behaviors. On the one hand, an audit firm will be motivated to behave non-independently for fear of losing the future stream of quasi-rents specific to a certain client[3]. On the other hand, the danger of being caught of behaving non-independently and thereby losing the stream of future quasi-rents specific to all other clients will serve to promote the independent behaviors. Mayhew and Pike (2004) and Menom (2003) reports that, when facing the countervailing incentives, larger firms are, in general, more independent and provide better monitoring.

Taking audit firm quality to be commensurate with the degree of auditor independence, the larger the audit firm, the higher the quality of its audit work. Prior research (e.g., Francis and Krishnan 1999) suggests that Big 4 auditors provide higher-quality audits in the US audit market in order to protect their firm reputation and to avoid costly litigation. In sum, despite some recent high-profile cases (e.g., Arthur Andersen's bankruptcy), the collective evidence is strongly supportive that audits conducted by large audit firms are of higher quality (Francis 2004; Watkins et al. 2004; Lee et al. 2003), or that the size of audit firm could be a useful proxy for the quality of audit services (Lennox 2005; Mansi et al. 2004).

Besides auditor size, some other dimensions are suggested to affect audit quality. It is reasonably well hypothesized that auditors who specialized in a given industry (industry specialists) have more opportunity to gain industry knowledge than their non-specialist counterparts. Schauer (2002) examined the association between bid-ask spreads[4], a proxy for audit quality, and auditors who specialize in their clients' industry. He found that firms audited by industry specialists had lower-levels of information asymmetry measured by bid-ask spread than firms audited by non-specialists. Since audits enhance the credibility of financial information, the reduction in information asymmetry associated with industry specialists is attributed to a higher level of audit quality in this group of audit engagements (Datar et al. 1991; Francis 2004).

The issue of audit quality is an area of immense importance for the development of China's stock market. DeFond et al. (2000) reported that the independence of auditing practice in China had improved, as evidenced by the increasing frequency of the modified opinions issued by Chinese auditors. In pace with the development of the stock market in China, a group of big auditors has also

---

[3] Quasi-rents are the excess of audit fees over the avoidable costs of performing audits. Quasi-rents arise due to the cost advantages of incumbency, which include the large startup costs associated with an audit engagement and the client-switching costs.
[4] A portion of bid-ask spreads arises from the difference in information asymmetry among stock market participants. So, the smaller the bid-ask spreads, the higher the audit quality.

emerged. DeFond et al. (2000) found that big auditors were more likely to issue the qualified opinions. Since the year of 2003, the CICPA has publicized the ranking of audit firms in China according to their annual revenues. Among all auditing firms in China, the average revenue of the ten largest auditing firms (Top 10) is substantially higher than that of others (non-Top 10). Since the audit quality is positively related to auditor size, Top 10 auditing firms should provide higher-quality audits in China. In this study, we use Top 10 to proxy for the quality auditors in China.

2.1.2 Corporate Governance, Opaqueness Gains, and the Monitoring Role of Audits

An agency relationship arises when a principal (e.g. an owner) engages another person as his/her agent to perform a service on his/her behalf. There thus is a problem of how to motivate the agent to act for the principal's interest rather than pursuing the agent's own benefit. The obvious answer is that the agent will not completely follow the principal's interest. There always exists the possibility that the agent will try to maximize self-benefits even at the expense of the principal's interest. As a result of information asymmetries between the principal and the agent and of their diverged motivations, the principal lacks trust in his/her agent and may therefore need to put in place some monitoring mechanisms, such as the audit, to reinforce this trust.

An audit provides an independent check on the work of the agent and on the information he/she provides, and therefore serves a fundamental purpose in assuring confidence and reinforcing trust in the contractual relationship between the principal and the agent. By performing the attestation function, auditors play a significant role in a firm's monitoring system. In this way, independent audits, as a necessary mechanism for proper functioning of contractual relations, are an essential component of the corporate governance mosaic. Hence, in principle, auditors must work with other actors in the corporate governance mosaic to ensure that management provides credible financial reports to stakeholders as well as help to protect the interests of current and potential investors.

An audit is valuable because it reduces the agency costs by assuring the credibility of client financial statements, thereby allowing more precise and efficient contracts to be based on the audited financial statements (Watts & Zimmerman 1986). Prior research argues that, as agency costs increase, a demand for higher-quality audits rises (e.g., Chow and Rice 1982; Watts & Zimmerman 1986; Francis & Wilson 1988). Independent audits serve as an important element of efficient equity markets, because the audits can enhance the credibility and quality of financial information and ultimately influence the allocation of resources in the market (Firth and Smith 1992; Ghosh and Moon 2005). Investors and creditors rely upon the audit results to evaluate business performance and make a variety of decisions (Feltham et al. 1991; Lin & Chen 2004).

Empirical studies indicate that the demand for audits as a corporate governance mechanism by firms in UK and US is a function of audit quality and the assurance provided by auditors (Willenborg 1999). Ashbaugh & Warfield (2003) argue that, as stakeholders demand for reliable financial information, external audits play a corporate governance role as a monitoring mechanism. The more independent the auditors who contribute to corporate governance of their clients, the more valuable their audits are (Hay and Davis 2004; Hudaib and Cooke 2005). Since high-quality auditors are more professional and more independent, they are more likely to discover and report irregularities and misstatements in the financial statements and therefore better serve as a monitoring mechanism (Lee et al. 2003).

In countries with poor protection of investors' rights, concentration of ownership often emerges as a solution to control moral hazard in managerial choices (La Porta et al. 1998; Shleifer and Vishny 1997). In such countries it is difficult to monitor enterprises at the arm's length and rely on market reactions to discipline the management (Perotti and Thadden 2005). The concentrated shareholding cannot by itself resolve conflicts among interested parties due to asymmetric information between the controlling shareholders and other investors, and may in fact aggravate the conflicts. For instance, the controlling shareholders may inefficiently redistribute wealth from other investors to themselves through "tunneling" behaviors (Shleifer and Vishny 1997).

High ownership concentration is a feature of the listed firms in China. At present there are three sources of equity capital for most listed firms in China, i.e., the state shares (represent the state's interest in a listed firm), the legal-entity shares (the equity held by state-owned-enterprises or social institutions) and the public shares (held by both institutional and individual investors). The state or a parent SOE is usually the largest shareholder and holds a significantly large proportion of the total equity. According to an investigation report, the share interest of the largest shareholder accounts for, on average, around 50% of the total equity of the Chinese listed firms in 2000. Under current 'split-share system', only public shares are tradable in the market. The state-shares and legal-entity shares are not tradable at present, so there is little incentive for the controlling large shareholders to care about the changes in the values or share prices of the listed firms in the market.

13

Concentrated ownership nevertheless induces agency problems. Monopolist control creates an entrenchment problem that allows controlling owners' self-dealings to go unchallenged internally by the BoD or externally by the takeover market. In addition, with a high level of ownership concentration, firms tend to disclose less information because of their expropriation and rent-seeking incentives. The controlling shareholders under this circumstance have incentives to keep corporate information private to derive the opaqueness gains (Fan and Wong 2002).

Fan and Wong (2002) also argue that low corporate transparency and less disclosure would lead to serious agency problems and poor corporate governance. Researchers have found that large shareholders may try to maximize self-utilities through 'tunneling' or benefit transfer, thus expropriate the interests of other shareholders and related parties (La Porta et al. 2002; Anderson et al. 2004; Hunton et al. 2006). This entrenchment problem can come at a price to the controlling owners and their firms: investors anticipate the problem; hence, they discount the share prices and raise the difficulty for the firms to issue equities in the future (Claessens et al. 2002; La Porta et al. 2002; Pittman and Fortin 2004).

Given the agency costs associated with a concentrated ownership, a controlling shareholder may introduce monitoring or bonding mechanisms that limit his/her ability to extract wealth from other shareholders and hence mitigate the agency conflicts. In this way, the controlling shareholder signals to the market that the firm is well governed and the interests of minority shareholders are sufficiently protected (Porter et al. 2003; Watkins et al. 2004). Using a broad sample from eight East Asian economies, Fan and Wong (2005) found that firms with agency problems embedded in the ownership structures were more likely to employ Big 5 auditors. The controlling shareholder may consider hiring a high-quality auditor will enhance the credibility of financial information disclosed to the market. Consistent with such an expectation, firms hiring Big 5 auditors receive smaller share price discounts associated with the agency conflicts. However, a monitoring mechanism will be adopted only when the benefits of imposing the mechanism (reduced agency costs or lowered capital raising costs) outweigh the costs of using the mechanism (e.g, the forfeited opaqueness gains due to the governance constraint). For instance, an auditor may inhibit a controlling owner's ability to manipulate earnings downward to justify the low cash dividends paid to outside shareholders.

Ashbaugh & Warfield (2003) documented that the demand for auditing as a corporate governance mechanism might be mitigated when equity shareholding is concentrated. A controlling owner may not desire to hire a high-quality auditor or he/she may even hire a low-quality auditor to reduce the external monitoring on his/her behaviors. Prior studies document that the weaknesses in corporate governance are often associated with low financial reporting quality, earnings manipulation and low level of corporate transparency (Beasley et al. 2000; Carcello and Neal 2000; Felo et al. 2001). The opaqueness (or low level of transparency) derived from weak corporate governance helps the controlling owner protect private benefits, even though at the expense of minority shareholders and other stakeholders.

Here we conceptually distinguish transparency from disclosure: transparency is a strategic ex ante decision, while disclosure is an ex post decision, since a firm may be reluctant to release unfavorable information ex post. Information dissemination, ex post, may include regular contacts with analysts, listing on an exchange with strict disclosure requirements, encouragement of active share trading, highly detailed annual reports, and a reputable auditor, and so on.

In China the benefit of opaqueness derived from weak corporate governance is even not limited to protecting private benefits. Political rent seeking is prevalent and highly lucrative in China. Firms in China may choose to disclose less to prevent competition or social sanctions. DeFond et al. (2000) found that as the Chinese government enacted regulations on domestic auditors to increase their independence, many of the listed firms took flight from high-quality auditors to low-quality ones, which gives evidence that Chinese listed firms would aggressively evade high-quality monitoring mechanisms in order to realize the opaqueness gains.

Fan and Wong (2005) found that, in eight East Asian economies, firms were more likely to appoint Big 5 auditors when their perceived entrenchment problems were more severe. Moreover, such relations are evident among frequent equity issuers but not among infrequent ones. Reed et al. (2000) reported that after Lavenhol and Horwath (LH) declared bankruptcy, LH clients that selected Big 6 auditors issued more securities after selecting the new auditor than LH clients that selected non-Big 6 auditors. Eichenseher and Shields (1986) also found that, comparing with non-equity issuing firms, the equity issuers demanded for higher-quality auditors. This suggests that when deciding whether to hire Big 5 auditors, the controlling owners would have to consider the trade-off between the benefits from raising capital and the costs of forfeiting the opaqueness gains.

Generally, audits can play a corporate governance role, but there are factors that raise doubt on this role in China. Many Chinese auditors were lack of expertise or willingness to supply quality audits (Yang

1995). In China, government control of the auditing profession appears to underpin the problems relating to auditor independence. At the national level, the profession is not self-regulated, but government-regulated. The professional oversight organization, the CICPA, is a government-controlled body with its senior officials appointed by the Ministry of Finance at the central government. The close association between CICPA and the government potentially reduces the independence of professional auditors (Yang et al. 2003). There are also some concerns that auditors' monitoring role may be in conflict with their consulting activities with client firms, an issue not unique to China. The enforcement of relevant laws governing the auditing profession is inadequate at present (Lin et al. 2006), which may have further undermined the independence of Chinese auditors.

2.2 Hypotheses Development

A listed firm in China usually has a controlling owner, being usually the government or a parent SOE. The controlling owner can exercise absolute control over the listed firm, including the auditor choice decision. The preceding discussion suggests that whether a firm hires a high-quality auditor to serve a corporate governance function is controversial, depending on the potential costs and benefits to the controlling owner. The state-shares and legal entity-shares are not tradable at present, so there is no motivation for the controlling large shareholders to care about the changes in the values or share prices of the listed firms in the market. In fact, the controlling shareholders of many listed firms, who are mainly government agencies or parent SOEs, are keen only in raising funds from the stock market (Xiang 1998).

For the Chinese listed firms, the main benefit of hiring a high-quality auditor is the firms may be able to raise funds in the capital market at a lower cost or sell shares at a higher price as the market may perceive that high-quality auditors will lead to better quality of information disclosure. And the costs will be the diminution of the opaqueness gains: the controlling owner may be inhibited in its ability to maximize self-interest through 'tunneling' or benefit transfer because of high-standard monitoring of the auditor. Basically, the controlling shareholders benefit from a lack of transparency: they engage in rent-seeking activities, at the cost of other shareholders (Leuz et al. 2001). In China, the controlling shareholders have frequently intervened in the operations of the listed firms to benefit parent companies, e.g., using the listed firms to guarantee loans for related entities, and exposing the listed firms to unnecessary financial and operating risks. They are frequently engaged in benefit transfer

16

through misappropriation of funds or related-party transactions to expropriate the listed firms and infringe upon the interests of other shareholders, the public investors in particular.

Our study on the Chinese stock market covers the period from the beginning of 2001 to the end of 2004. After the establishment of Shanghai Stock Exchange and Shenzhen Stock Exchange in the early 1990s, the Chinese listed firms have achieved an accumulated financing amount of RMB 1.16 trillion between 1992 and 2004, and the total market capitalization once hit RMB 1.61 trillion in 2000. But the five-year bear market since 2000 has resulted in a market value slump by RMB 0.44 trillion to RMB 1.17 trillion (the CSRC 2005). Because of the weak market, the listed firms have trivial intentions of offering new equity to the public from 2001 to 2004. CSRC even suspended the Chinese listed firms from issuing new equity to the public in June 2002. Nevertheless, IPO equity offerings were still market-widely oversubscribed in the market because IPO prices are usually set at a comparatively low level. The stock market saw immediate quick rise of IPO equity prices, owing to relatively little supply in the stock market.

In such a market, the benefits of lowering capital raising costs are of insignificance: the Chinese listed firms have little intention or possibility of offering new equity to the public; IPO firms can sell their shares with no difficulty, but they are unable to offer new equity in the near future after listing. Therefore, the opaqueness gains are supposed to outweigh the benefits of lowering the costs of capital raising. Hence, lower-quality auditors will be preferred by the listed firms, especially by the listed firms with weaker internal corporate governance mechanism, because they have relatively more opaqueness gains to protect (Beasley et al. 2000; Carcello and Neal 2000; Felo et al. 2001).

To address this issue empirically, we intend to test the relation between firms' decisions on auditor choice and their internal corporate governance mechanism: whether firms with weaker internal corporate governance mechanism, and therefore potentially with higher opaqueness gains, would less likely choose Top 10 auditors, who are generally perceived to be more independent and of higher quality.

The internal corporate governance mechanism within a firm consists of various types of organizational arrangements or procedures to balance the power and responsibilities among the firm's shareholders, directors, the management and the employees. Among them, the ownership structure, the board of directors (BoD), the supervisory board (SB) and the duality of the BoD chairman and the CEO are of

17

great importance in determining the effectiveness of internal corporate governance mechanism, especially for the listed firms in China (Liu and Sun 2005; Sheng 2004).

Ownership structures affect corporate governance and corporate value in many complex ways. Daniels and Iacobucci (1999) argue that more narrowly held firms may face greater agency costs because the controlling shareholders would have a dominant influence on corporate affairs and it is easier for them to bypass the monitoring of other shareholders. La Porta et al. (1998, 1999) showed that in the emerging transitional economies, the controlling shareholders may expropriate the minority shareholders through aggressive "tunneling" behaviors. They further argue that "the central agency problem in large corporations around the world is that of restricting expropriation of minority shareholders by controlling shareholders" (La Porta et al. 1999). This is particularly true for the Chinese listed firms where the controlling shareholders, on average, hold a very large portion of the equity.

There are two primary patterns of board structure: the unitary-board system or Anglo-American system, and the two-board system or the German-Japan system. Under the unitary-board system, a company has only one board, comprised of the executive directors and the independent directors. The executive directors are in charge of the company's business operation, while the independent directors act as supervisors of the management. Under the two-board system, a firm has two boards --- the BoD and the supervisory board (SB). The SB functions as the special monitoring organ and may have the same mandates as the BoD or even higher status than the BoD. Pursuant to the Chinese Company Law, the listed firms adopt the typical two-board system, thus each listed firm has both the BoD and the SB.

According to the Company Law, the supervisory board shall be composed of shareholders' representatives and an appropriate proportion of employee representatives who are nominated by the employee union of the firm. The Code of Corporate Governance for Listed Companies in China issued by the CSRC and State Economic and Trade Commission (2001) further requires that members of SB should have professional knowledge or work experience in such areas as business law and accounting. The SB shall ensure its capability to independently and effectively conduct its supervision over the activities carried out by the directors and the management as well as to monitor or examine the financial affairs of the firm.

The board of directors (BoD) must ensure that the management acts in the interests of the shareholders.

The BoD is responsible for the execution of the resolutions passed in the shareholders' meetings; for appointing, removing and remunerating the general managers and other senior managers. However, many directors are concurrently the executives of the firm (including the CEO). As a result, they are less likely to be impartial in supervising and evaluating the performance of the management. For the BoD to effectively perform a monitoring function, the separation of the positions of the CEO and the BoD chairman is essential in respect of an effective internal corporate governance mechanism (La Porta et al. 1999).

We use three proxies to measure a firm's internal corporate governance mechanism: the degree of ownership concentration (shareholding of the largest owner), the SB size, and the duality of the CEO and the BoD chairman. There are other variables appropriate for measuring internal corporate governance mechanism, such as the function of independent directors and audit committees. These internal monitoring variables are not included in our model because they were introduced into China only after 2002 and their full adoption is out of the test period.

Extracting private control benefits, if detected, is likely to invite external intervention by minority shareholders, analysts, stock exchanges, or market regulators (Haw et al. 2004). The desire to maximize self-interest through 'tunneling' or benefit transfer may drive the listed firms to avoid being monitored by Top 10 auditors. In general, the more concentrated the ownership structure, the weaker the internal corporate governance, hence, there will be more opaqueness gains for the controlling shareholders (Chau and Leung 2006). Therefore firms with larger controlling owners are more eager to choose a pliable auditor so that they can easily benefit from tunneling behaviors under a lower level of audit monitoring. In addition, the controlling shareholders have the monopolist position and influence so they can easily control or dominate the nomination and appointment of directors and senior management officers and virtually preclude other shareholders from participation in making operating decisions, including hiring of auditors. Hence, a larger controlling shareholder is more likely to choose a low-quality auditor in order to realize and sustain the opaqueness gains.

Hypothesis 1 (H1): ceteris paribus, the higher percentage of total shares held by the largest owner, the less likely a Top 10 auditor will be chosen.

Under the requirement of the Company Law, Chinese firms adopt a dual-board system, including the board of directors (BoD), and the supervisory board (SB). China has adopted this system since 1994 for the publicly listed firms. It is required that SB members hold neither the position of a board director, nor CEO or chief finance officer. The Company Law specifically defines the SB as a monitoring mechanism, carrying out a series of responsibilities, including: (1) monitoring the performance of directors and managers, to ensure compliance with laws, regulations, and the articles of incorporation, (2) reviewing the financial affairs of the company, (3) requesting directors and managers to alter and/or rectify their personal actions if they are in conflict with the firm's objectives, (4) proposing temporary shareholder meetings whenever they deem necessary, (5) fulfilling any other duties that are stipulated in the articles of incorporation of the firm, and (6) submitting a SB report to the shareholders' annual general meeting.

It is argued that the dual-board structure may have additional advantages over a unitary-board structure, particularly in China where the external monitoring devices, such as the market regulation and surveillance systems have not been well developed. The SB is concerned with the firm's strategy and stakeholder interests, but does not interfere in the operational management. So, the existence of SB would provide a basis for managerial autonomy, as well as for monitoring by stakeholders (Chen 2005).

The SB is entitled to exercise supervision over the work of the BoD, the performance of the management and the business financial affairs. Such German-styled two-tier board system with co-existence of the BoD and the SB, in fact, has become the backbone of corporate governance in the Chinese listed firms since the mid 1990s. Using an event study, Dahya et al. (2003) found that investors considered the SB to be important in corporate governance in China. Chen (2005) found that there is a positive association between the size of SB and the level of corporate governance. More members in a SB will enhance its monitoring role. Hence, we use the number of SB members to proxy for the monitoring strength of the SB. The fewer SB members a listed firm has, the weaker the internal corporate governance is, hence the more opaqueness gains will there be. As a result, the controlling shareholders in firms with a small SB size will be more eager to choose lower-quality auditors to protect their opaqueness gains.

Hypothesis 2 (H2): ceteris paribus, a firm with fewer SB members is less likely to choose a Top 10 auditor.

Traditionally, in American businesses, the same person occupies the positions of the CEO and the BoD chairman. In most European and Canadian businesses, duality of the two positions is required, in an effort to ensure better governance of the firm (i.e., maintaining a balance of power between the two critical positions). Combining the roles of these two positions does have its advantages, giving the CEO multiple perspectives on the company and empowering him/her to act with determination. However, this allows for little transparency of the CEO's acts, and as such his/her actions can go unmonitored since the CEO will have a dominant influence in the BoD meetings as he/she serves as the BoD chairman concurrently. Non-separation of the two positions may also lead to corporate scandals and corruptions as there is a lack of internal checking in the two most powerful positions in the corporate decision-making process. An effectively independent board is the shareholders' best protection. Duality of the two positions allows the chairman to monitor the work of the CEO, and in turn the company's overall performance, on behalf of the stockholders. Ultimately, when the BoD chairman does not occupy the position of the CEO, he will be able to govern the firm in a more impartial manner (Cohen et al. 2002; Gelb and Zarowin 2002).

Investors, researchers, and government officials gradually accept the view that best practices of corporate governance require the separation of the roles of the BoD chairman and the CEO and the installation of a power balancing and checking mechanism. The idea that the duality of the CEO and the BoD chairman positions will provide an advantage to investors is based on the belief that a well-designed corporate governance system can reduce the agency costs. The separation of the CEO and the BoD chairman positions received a boost in 2003 after recent corporate scandals. It is considered of bad corporate governance if the same person takes both positions of the CEO and the BoD chairman, which may lead to less transparency of, and weak monitoring device within, the firm (Imhoff 2003; La Porta et al. 1999).

Hypothesis 3 (H3): ceteris paribus, a firm with no duality of the positions for CEO and BoD chairman is less likely to choose a Top 10 auditor.

2.3 Research Methodology

2.3.1 Model Specification

In this study we use a binary classification to divide auditors in China into two categories: the ten largest auditors (Top 10) to proxy for high-quality auditors and non-Top 10 auditors to proxy for low-quality ones. As discussed previously, audit firm size is an effective surrogate for audit quality

21

(DeAngelo 1981; Lee et al. 2003; Lennox 1999). We construct a model to examine whether firms' auditor choice decision is associated with their internal corporate governance mechanism. If the two types of auditors provide a similar level of monitoring, then firms would randomly select auditors, suggesting that internal corporate governance mechanism should not dominate the choice of auditors. In contrast, if the different types of auditors offer monitoring services with varied quality, then the firms' internal corporate governance mechanism should predict their choice of auditors, in terms of the varied motivations and monitoring function on their opaqueness gains.

The following auditor choice model (Equation (1)) is run by logit regression to test Hypotheses 1 to 3.

$$AUD = \beta_0 + \beta_1 LSH + \beta_2 SB + \beta_3 CEOCHR + \beta_4 LNASSET + \beta_5 ATR + \beta_6 ROA +$$
$$\beta_7 CURR + \beta_8 DA + \beta_9 ABBETA + \varepsilon \qquad (1)$$

Where:

AUD	=1 if the auditor is a Top 10 auditor; 0 otherwise
LSH	=the largest owner's shareholding as a percentage of total shares
SB	=number of SB members
CEOCHR	=1 if the CEO also holds the position of BoD chairman; 0 otherwise
LNASSET	=log of total assets
ATR	=asset turnover ratio, calculated as sales divided by total assets
ROA	=earnings after interest and taxation divided by total assets
CURR	=current assets divided by total assets
DA	=total liabilities divided by total assets
ABBETA	=the absolute value of beta to proxy for firm risk

We control for firm size, assets turnover ratio (growth), profitability, assets structure, financial leverage, and firm risk in the model. Willenborg (1999) suggested that large IPOs could be forced to hire high-quality auditors under the terms of underwriting engagements. Large IPOs are usually more complicated in operation, and therefore need to hire auditors with more expertise, which are usually large auditors. Economies of scale also increase the probability that large IPOs select a high-quality auditor, as the high-quality auditing firms (usually of large size) are able to audit large IPOs at low average costs (Chaney et al 2004). Hiring a high-quality auditor may also provide more incremental proceedings to the large IPO firms as a high-quality audit enhances the stock price. We use log of total assets to control the size effect of the IPO firms.

Anderson et al. (2004) reported that firms with high assets turnover ratio were inclined to choose high-quality auditors. Willenborg (1999) documented that firms audited by large auditors were more profitable, ceteris paribus. For one side, more profitable firms are more eager to hire a high-quality auditor to testify their performance to the market. In addition, more profitable firms usually have more money to hire a large or a high-quality auditor.

Reed et al. (2000) found that firms selecting Big 6 auditors tended to be more highly leveraged. They argue that more leveraged firms have stronger incentives to hire a high-quality auditor to reduce the market's suspicion on their performance. In contrast, Titman and Truman (1986) predicted that entrepreneurs of high-leveraged firms were more likely to choose lower-quality auditors. Their prediction follows from the argument that the effect of information on the market value of an IPO increases with the precision of the information. Thus, a firm with bad information (e.g., higher financial leverage) will find out less benefit in hiring a high-quality auditor. Other studies also suggest that the assets structure and financial leverage are related with firms' auditor selection decisions, but the direction of the relationship is inconsistent (Wallace 1987; Friedlan 1994; Firth and Smith 1992).

As IPO firm-specific risk increases, so does the signaling value associated with the utility of a high-quality audit firm. Willenborg (1999) argues that high-quality auditors can provide a source of "assurance" to IPO investors: if the audited financial information turns out to be misleading, investors can sue auditors for damage award. The value of such assurance is increasing in respect of the degree of firm-specific risk. However, empirical results are not generally consistent with this prediction. Tests using US IPOs (Simunic and Stein 1987; Beatty 1989; Feltham et al. 1991) reveal that firm-specific risk and choice of quality audits are negatively related. Copley and Douthett (2002) however, found that the demand for large audit firms increased with the extent of firm-specific risk. The model run by Datar et al. (1991) also suggests that the incentive to engage a high-quality auditor increases with the degree of risk. While auditor choice may not perfectly reveal a firm's private information, it does

23

reduce the level of uncertainty. Hence, riskier firms may benefit from hiring a high-quality auditor because by hiring a high-quality auditor, their information uncertainty can be reduced to a greater extent than less risky ones.

In this study, the absolute value of beta is used to proxy for the risk of IPO firms. Beta is commonly used to proxy for firm-specific risk in finance literature. According to its definition, beta represents the relative stock price volatility to the market. Stocks of IPO firms are not traded until they get listed, so we use the daily stock prices 3 months after the IPO date to derive beta. Different from prior research, we use the absolute value of beta, rather than beta itself, to control for the risk factor. According to relevant finance theory, beta can be either positive or negative (though negative values are less frequent). If beta for a firm is positive, stock prices of that firm move in the same direction with the market; and if beta is negative, stock prices move in the opposite direction with the market. Signs of beta measure the relative direction of stock price movement in respect of the market movement, and only the absolute value of beta measures the volatility of stock prices, which can be a proxy for the degree of risk.

2.3.2 Sampling

Our sample covers the A-share firms getting listed at Shanghai and Shenzhen Stock Exchanges from the beginning of 2001 to the end of 2004. There are two main reasons for using firms listed during this time period. The first one is about the classification of high-quality auditors in China. The CICPA started to publicize the ranking of audit firms in China according to their annual revenues in 2002. So far, CICPA has publicized the ranking of Chinese audit firms for three years, from 2002 to 2004. Based on the average revenues for the three years, the ten largest auditors (Top 10) in China can be sequentially identified as: PwC Zhongtian, KPMG Huazhen, Deloitte Huayong, EY Huaming, Lixin Changjiang, Yuehua, Xinyongzhonghe, Beijing Jingdu, Jiangsu Gongzheng, and EY Dahua [5]. Theoretically, Top 10 auditors are subject to change every year, but during our study period the rankings are fairly stable. These ten auditors are ranked among the largest twelve auditors for each of the three years, with only EY Dahua ranked 17th for 2004. Considering the stableness of rankings, it

---

[5] According to the rule of CICPA, a foreign accounting firm is allowed to conduct business in China only through setting up a joint venture with a local firm, for example, EY Dahua is a joint venture set up by Ernst & Young and a local Chinese firm Dahua CPAs.

can be assumed that they represent the high-quality auditors in China. To expand our sample size, we assume that this classification of high and low-quality auditors also holds for 2001.

[insert Table 1 here]

The second main concern for sampling is the listed firms' rights offering intentions. For the controlling shareholders of the Chinese listed firms, the main concern is to raise money in the stock market. The controlling shareholders have an incentive to adopt effective external monitoring measures to reduce agency costs so that they can raise money at a lower cost. However, they would also try to avoid stringent external monitoring so as to protect the opaqueness gains. When the possibility of rights offering is low, the opaqueness gains outweigh the benefits from low capital raising costs. China has experienced five-year bear market since 2000. In such a market, the listed firms have remote intentions and possibilities of offering rights and new stocks because their stock prices are generally low. Therefore during 2001 to 2004, the opaqueness gains derived from weak corporate governance mechanism outweigh the benefits from lowering capital raising costs. Hence, the motivation for opaqueness gains dominates firms' auditor choice decisions. The period from 2001 to 2004 is thus appropriate to test the relationship between the auditor choice decisions and the internal corporate governance mechanism of the listed firms.

Data are collected from CSMAR Financial Database and CSMAR IPOs Research Database published by GTA Information Technology Company (China), which are the most commonly used database for the market-based research on the Chinese market at present. To ensure their reliability, the collected data are double checked with TEJ Database and the authoritative newspapers and magazines designated by the CSRC, such as China Security Daily, Shenzhen Security Times, and Shanghai Security Daily. Financial, transportation, and utility firms are excluded from this study because their operations and governance structures are quite different from all other types of firms in nature. Description of the data used in our study is provided in Table 2.

[insert Table 2 here]

Panel B of Table 2 presents the basic descriptive statistics of variables. The audit market in China is somewhat different from that in the US. Big 4 dominate the market and audit most of the listed firms in the US. However, Top 10 in China have audited only 22.8% (42/184) of the sample IPO firms in this study. The largest owners can usually have absolute controlling power over the listed firms, holding a mean of 46.09% (median of 48.83%) of the total shares of the sample IPO firms. This phenomenon of monopolist controlling shareholding has actually led to poor corporate governance practices and raised serious concerns of investors, professionals, regulators, and the public. For instance, the controlling shareholders frequently exercised their controlling powers to exploit the interests of minority shareholders and virtually preclude other shareholders from participation in making business decisions (Lin et al. 2006). Hence, tunneling behavior is extremely obvious in China. For the sample firms, ROA is reasonably good with a mean of 10.2%. However, we need to maintain a skeptic attitude as IPO firms often do a lot of financial packaging before public listing. Overall, the debt-to-asset ratio is very high with a mean of 53.94%, suggesting a low-level of financial health for the Chinese listed firms.

Panel C of Table 2 presents the correlation coefficient matrix. The choice of Top 10 auditors is positively related at the significant level, with the SB size, log of total assets (firm size), assets turnover ratio (growth) and debt-to-asset ratio (financial leverage), and significantly negatively related with shareholding of the largest owners (ownership concentration), and concurrent appointment of the CEO and the BoD chairman. Correlation coefficients among the independent variables are not high with only one at the level of -0.666 (between the current ratio and the debt-to-asset ratio), so the multicollinearity problem is moderate and won't have a significant effect on the relation between the dependent and independent variables in Model (1).

2.4 Empirical Results

Table 3 provides empirical test results from the regression to identify whether the opaqueness gains derived from weak internal corporate governance mechanism preclude firms to choose high-quality auditors. As the dependent variable (audit choice) is binary in nature (Top 10 or non-Top 10), we use the logit regression. Our sample is the IPO firms getting listed at Shanghai and Shenzhen Stock Exchanges during 2001 to 2004. After deleting firms with missing data and the financial, transportation, and utility firms, the final sample consists of 184 IPO firms. With a pseudo R-square of 0.331 and a Chi-square of 45.233, the model is significant and can differentiate the listed firms choosing Top 10 auditors from those choosing non-Top 10 auditors.

The coefficient for the variable of ownership concentration (the largest owner's shareholding) is negatively significant at 1% level (β_1=-0.057, Wald=19.018). It suggests that with larger controlling owners, the listed firms are less likely to choose a Top 10 auditor. The finding is consistent with H1.

The SB size is positively related to the selection of a Top 10 auditor at 5% level (β_2=0.268, Wald=5.671), suggesting that firms with larger SB size are more likely to choose a Top-10 auditor, which is supportive to H2. Consistent with H3, CEOCHR is negatively related with choosing a Top 10 auditor, but only at a marginal level of 10% (β_3=-1.589, Wald=3.311), suggesting that when the CEO also holds the position of BoD chairman, the listed firm is less likely to hire a Top 10 auditor.

For the control variables, firm size (log of total assets), growth (assets turnover ratio), and profitability (return on assets) are all positively related to the selection of a Top 10 auditor. Consistent with prior research, the results suggest that larger firms, firms with higher growing potential, and more profitable firms are inclined to hire Top 10 auditors. The coefficients for liquidity (current-to-total-assets ratio), financial leverage (debt-to-assets ratio), and financial risk (absolute value of beta) are not significant.

[insert Table 3 here]

In summary, the empirical results support the three hypotheses. During the period 2001 to 2004, there was a continuous retreat in the stock market in China. In such a bear market, the listed firms have little intention of offering new stocks and rights to the public. Also, the listed firms in China are usually not allowed to issue bonds and their ability to borrow money from banks is restricted at present. Hence, for the controlling owners of the listed firms, the potential opaqueness gains outweigh the benefit of lowering capital costs. As firms with weaker internal corporate governance mechanism have more opaqueness gains, they are more likely to avoid Top 10 auditors.

2.5 Sensitivity Tests

A series of sensitivity tests were conducted to examine the robustness of the model and the empirical results. Although most listed firms in China have very concentrated ownership structure, some do not. In order to make sure that the largest shareholders control the listed firms and are able to effectively influence the auditor choice decisions, we limit the sample to firms in which the largest owners hold a significant percentage of the total equity shares. We delete the observations in which the largest owners hold less than 10% of the total equity shares, and the empirical results are substantially the same (Table 3-1).

27

We also applied alternative proxies to measure the control variables. We used the log of revenues to proxy for firm size, the return on equity to proxy for profitability, the long-term debt to asset ratio to proxy for financial leverage, and the Z-score to proxy for firm risk. H1 to H3 are generally supported after adopting the alternative measures for the control variables. We also reran the regression after deleting observations that are more than three standard deviations away from the mean and the results remained substantially the same.

Chapter Three

The Impact of Internal Corporate Governance on Auditor Switch

3.1 Hypothesis Development

Another important and related research question is when the listed firms switch auditors, whether the firms with weak internal corporate governance mechanism are more likely to switch to an auditor of lower quality. As argued before, during 2001 to 2004, the motivation for the opaqueness gains tend to outweigh the benefits of lowering capital raising costs for the Chinese listed firms. So lower-quality auditors will relatively be preferred by firms with weak internal corporate governance mechanism because the controlling shareholders of those firms have more opaqueness gains to protect.

The "tunneling behavior" is extremely obvious in China due to the unique equity structure of the Chinese listed companies. At present only the public shares are freely tradable on the stock exchanges. The controlling shareholders of the listed firms are usually the government or parent SOEs and their shares are not freely tradable. As a result, the controlling owners are less enthusiastic about stock prices and they would obtain shareholding benefits mainly through tunneling behaviors.

In this chapter, we test the relationship between the firms' internal corporate governance mechanism and the relative quality of their successor auditors, namely whether the quality of a successor auditor will be higher or lower than the predecessor auditor when auditor switch is made. As audit quality is difficult to observe, we use auditor size to proxy for the audit quality, which is consistent with prior research (DeAngelo 1981; Francis and Krishnan 1999; Francis 2004; Watkins et al. 2004; Lee et al. 2003). Using the same three proxies for the internal corporate governance mechanism, we set the following three hypotheses:

Hypothesis 4 (H4): ceteris paribus, the higher percentage of total shares held by the controlling owner, the more likely the firm will switch to a smaller auditor.

Hypothesis 5 (H5): ceteris paribus, a firm with a smaller SB size is more likely to switch to a smaller auditor.

Hypothesis 6 (H6): ceteris paribus, a firm with the CEO and the BoD chairman held by the same person is more likely to switch to a smaller auditor.

3.2 Research Methodology

3.2.1 Model Specification

Among firms that switch auditors only once during 2001-2004, we divide them into two types: those switching to a larger auditor (SU firms) and those switching to a smaller auditor (SD firms) (The whole ranking of auditors in China is provided in Appendix). As illustrated previously, audit firm size is an effective surrogate for the independence and monitoring strength of auditors in the literature (Francis and Krishnan 1999; Watkins et al. 2004; Lee et al. 2003). Switching to a smaller auditor may facilitate the listed firms to avoid more stringent monitoring from their predecessor auditor. As firms with weak internal corporate governance mechanism have more opaqueness gains to protect, they are more likely to switch to a smaller auditor to avoid rigorous monitoring. We construct a model to examine whether the direction of auditor switch (namely switching to a larger auditor or a smaller one) is associated with the firms' internal corporate governance mechanism (with the concentration of ownership, SB size and duality of the CEO and the BoD chairman positions as the proxies).

We run logit regression on the following auditor switch model (Equation (2)) to test Hypotheses 4 to 6.

$$SD = \gamma_0 + \gamma_1 LSH + \gamma_2 SB + \gamma_3 CEOCHR + \gamma_4 LNASSET + \gamma_5 LEV + \gamma_6 MB + \gamma_7 LOSS + \gamma_8 OPI + \varepsilon \tag{2}$$

Where:

SD	=1 if the firm switches from a Top 10 to a non-Top 10 auditor; 0 if the firm switches from a non-Top 10 auditor to a Top 10 auditor
LSH	=the largest owner's shareholding as a percentage of total shares
SB	=number of SB members
CEOCHR	=1 if the CEO also holds the position of the BoD chairman; 0 otherwise

LNASSET	=log of total assets at the end of the previous year
LEV	=long-term liabilities divided by total assets at the end of the previous year
MB	=market-to-book ratio at the end of the previous year, calculated as the market value of stocks divided by the book value
LOSS	=1 if the firm experiences a net loss for the previous year; 0 otherwise
OPI	=1 if the firm receives an unclean auditor opinion for the previous year; 0 otherwise

Firms may have very different reasons for switching to different types of auditors (e.g., switching to a larger auditor to improve corporate governance and company image or switching to a smaller auditor to sustain the opaqueness gains). In addition to the three corporate governance variables, the effects of some firm-specific factors that are likely to affect a firm's switch type are controlled in the regression, including firm size, financial leverage, growth, profitability, and auditor's opinion. There are some studies on auditor switch in the literature, but they do not probe into the types of auditor switch (e.g., switching to higher or lower-quality auditors). Nonetheless, the basic theories and findings in prior research on auditor switch can be applied to our study on the types of auditor switch.

Larger firms have smaller incentives to switch to a smaller auditor, since the financial analysts and the financial press will scrutinize their auditor switch more closely. In addition, a small auditor may not be professionally competent to audit a large firm. Hence we expect larger clients to be less likely to switch to a smaller auditor, which is also consistent with prior findings (Beattie and Fearnley 1995; Carcello and Neal 2003). Craswell (1988) reported that firms in financial distress might have greater incentives and possibilities to change auditors. He argued that firms in financial distress might have more conflicts with their auditors and would therefore try to switch to more pliable auditors. Eichenseher and Shields (1986) and Francis and Wilson (1988) documented a relationship between client's leverage level and changes to large auditors. We also include the market-to-book ratio in the model to control for the propensity of growing firms to switch to less conservative auditors, which is found by DeFond and Subramanyam (1998). Johnson and Lys (1990) found that firms switching auditors were less profitable than those that did not switch. Less profitable firms usually have more incentives to manage earnings

31

and therefore may have more conflicts with their auditors (Lee et al. 2004). Hence we control the factors of firm size, financial leverage, growth and profitability in the regression.

Another very common reason cited for auditor switching is the qualification of auditor opinions. Prior research finds that clients receiving an unclean auditor report are more likely to switch auditors (e.g. Chow and Rice 1982; Geiger et al. 1998), perhaps because the management believes that once an incumbent auditor is dismissed, the company will be able to find a more pliable auditor to obtain a clean report (Craswell 1988). Alternatively, the management might dismiss an auditor solely as a punishment for issuing a going-concern report, or due to irreparable damage to its relationship with the auditor as a result of the conflict. Auditor's opinion is likely to be related to the type of auditor switch and is thus incorporated in the model.

3.2.2 Sampling

Our sample covers A-share firms that switched auditors from the beginning of 2001 to the end of 2004. Consistent with the auditor choice study in the previous chapter, there are two main reasons to limit the sample firms that have switched auditors during this time period. The first one is the availability of the ranking of auditors. And the second one is that firms should have little intention to offer the rights or new stocks to the public during the time period, therefore the opaqueness gains derived from weak internal corporate governance mechanism significantly outweigh the benefits from lowering capital raising costs. The period from 2001 to 2004 is hence appropriate to test the relationship between auditor switch and the internal corporate governance mechanism. Data are collected from CSMAR Financial Database, TEJ, China Security Daily, Shenzhen Security Times, and Shanghai Security News. Financial, transportation, and utility firms are excluded from our study because they are, in nature, very different from all other types of firms. Description of the data used in our regression model is provided in Table 4.

[insert Table 4 here]

At the end of 2004, there are 1,387 A-share listed firms in China, among which 316 firms (22.7%) switched auditors during the four year period of 2001-2004. This implies that, generally, firms are not willing to switch auditors because of the high switching costs[6]. Panel A of Table 4 demonstrates the sample size of our study. As explained in the previous chapter, we exclude the financial, transportation and utility firms. We also delete firms that switch auditors for more than once during the four-year period. Firms switching auditors frequently may indicate serious underlying reasons and those reasons are out of the scope of our study. Also, firms switching twice or more in the period may switch to a larger auditor once and to a smaller auditor for another time, making it difficult to categorize their switch types. The final sample consists of 62 firms.

Panel B of Table 4 lists basic descriptive statistics of the variables incorporated in Model (2). Among the 62 sample firms, 36 firms switched to larger auditors, while 26 firms switched to smaller auditors during the period. The controlling owners of these firms, on average, hold 47.82% of the total shares, indicating the high ownership concentration of the Chinese listed firms. In the sample, 16.1% (10/62) of the firms receive unclean auditor opinions before their switch of auditors. This high percentage may indicate an association between auditor switch and unclean auditor opinions received.

Panel C of Table 4 presents the correlation coefficient matrix of the variables run in the regression. Firms' switching to a smaller auditor is, at a significant level, positively correlated to the largest owner's shareholding, the size of SB, and the CEO's holding the position of BoD chairman, and is negatively related to firm size (log of total assets) at a significant level. Correlation coefficients among independent variables are moderate with the largest one being 0.598 (between whether a firm experiences a net loss and whether it receives an unclean auditor opinion), so the multicollinearity problem is not serious and won't have a significant effect on the relation between the dependent and independent variables.

3.3 Empirical Results

Table 5 outlines the empirical results from the regression which tests whether, among switching firms, the firms with weaker internal corporate governance mechanism are inclined to switch to a smaller auditor. Since the independent variable has binary value (1 for switching from a Top 10 to a non-Top 10 auditor, and 0 for switching from a non-Top 10 to a Top 10 auditor), the Ordinary Least Square

---

[6] There are high costs for a listed firm to switch its auditor, for example, the firm needs to incur negotiation costs, the new auditor needs time and efforts to get familiar with the firm's operation, and investors may respond negatively to auditor switch (Teoh 1992).

(OLS) method does not work for the model. Rather, we use logit regression. With a Chi-square of 19.576 and a pseudo R-square of 0.364, the model is satisfactory at a significant level and can well differentiate firms switching to a smaller auditor from those switching to a larger one.

The coefficient for the largest owner's shareholding (concentration of ownership) is positively significant at 5% level (γ_1=0.046, Wald=4.467), which supports H4. It suggests that firms with larger controlling owners are more likely to switch to a smaller auditor. The coefficient for SB size is insignificant (γ_2=0.207, Wald=1.498), therefore H5 is not supported, indicating that the size of SB is not significantly related to the types of auditor switch. Consistent with H6, the CEO's holding the position of BoD chairman is positively related to the switching to a smaller auditor at 5% significance level (γ_3=2.955, Wald=5.133). So, a firm is more likely to switch to a smaller auditor if the duality of the CEO and the BoD chairman positions is not maintained.

Consistent with our predictions, firm size is negatively associated with the switching to a smaller auditor, while financial leverage has a significantly positive coefficient. Hence, large firms are less likely to switch to a smaller auditor, while firms with high financial leverage are inclined to do so. The coefficients for MB, LOSS and OPI are insignificant, suggesting that whether a firm is fast in growth, experiences a net loss or receives an unclean auditor report does not have a significant effect on its auditor switch types. The reason why auditor opinions do not have a significant influence might be that we did not separate different types of unclean opinions (i.e., unqualified with explanation, qualified, adverse, and disclaimer).

[insert Table 5 here]

In summary, the empirical results support that there is an association between the internal corporate governance mechanism and firms' auditor switch decision. The weaker the internal corporate governance mechanism is, the more likely the firm will switch to a smaller auditor rather than to a larger one. The result may further suggest that the opaqueness gains derived from weak corporate governance mechanism might be a consideration when firms switch auditors. For the sake of protecting their opaqueness gains, firms with weaker internal corporate governance mechanism are inclined to switch to a smaller, or more pliable, auditor.

3.4 Sensitivity Tests

To examine the robustness of the model and the empirical results, we run different sensitivity tests. First, we extend the proxy for SD and SU to include more observations. In the main test, SD is defined as switching from a Top 10 auditor to a non-Top 10 auditor and SU as switching from a non-Top 10 auditor to a Top 10 auditor. In the sensitivity test, SD and SU are not limited to switches between a Top 10 and a non-Top 10 auditor and also include switches among Top 10 auditors and switches among non-Top 10 auditors. For example, switching from 30th to 20th auditor is considered a switch from a smaller auditor to a larger one, while switching from 3rd to 8th auditor is considered a switch from a larger auditor to a smaller one, and so forth. In this way, the sample size is greatly enlarged to 233 observations. The empirical results remain unchanged (see Table 5-1). In order to make sure that the largest shareholder does control the listed firm, we limit the sample to firms in which the largest owners own a significant percentage of the total shares. After deleting the observations in which the largest owners hold less than 10% of the total shares (actually among the 62 observations there is only one such case), the empirical results remain substantially the same.

We also apply alternative proxies to measure the control variables. We use log of revenues to proxy for firm size, total debt to asset ratio to proxy for financial leverage, and increase of assets for growth. After adopting different measures for these control variables, the empirical results still demonstrate a positive relation between SD (switching to a smaller auditor) and LSH (largest owner's shareholding percentage) and CEOCHR (the CEO's holding of the BoD chairman position) at a significant level. However, the coefficient for SB (the size of SB) remains insignificant. Although the Company Law empowers the SB to monitor the work of BoD, the management, and the financial affair of the firm, it does not prescribe how the SB can exercise the power as well as the liabilities of SB members in case of breach of duty. Therefore, the supervisory function of the SB in terms of monitoring the management and corporate affairs is very limited. That the coefficient for SB remains insignificant in the test model may reflect that the SB could not perform a corporate governance role satisfactorily in China. Some researchers claim that the SB is mainly a decorative setting because its members are mostly from inside the firm and cannot function effectively in terms of its monitoring role (Xiao et al. 2004). Our study findings confirm such an assertion.

We also rerun the regression after deleting the observations that are more than three standard deviations away from the mean and the results remain unchanged. Therefore, H4 and H6 are robust to sensitivity tests, but H5 is not supported.

Chapter Four

The Market Implications of Auditor Choice and Auditor Switch

4.1 Literature Review

4.1.1 Information Quality, Information Credibility and Audit Quality

Information quality refers to how well the financial statements reflect the true economic conditions of the company. Francis and Krishnan (1999) argue that the management is less (more) able to manipulate earnings to meet the forecasted earnings when they hire auditors with high (low) quality. Prior studies provide support for an inverse association between the auditor brand name and the propensity for earnings management. Becker et al (1998) found that discretionary accruals were higher for firms with non-brand name auditors than for firms with brand name auditors because firms employing higher-quality auditors would have less opportunity to use accruals to manage earnings.

Krishnan (2003) defines information quality in terms of its ability to predict future profitability. His findings suggest that the discretionary accruals of firms with brand name auditors are a better predictor of future earnings and future cash flows than the discretionary accruals reported by firms with non-brand name auditors. His study supports prior theoretical suppositions that brand name audit firms offer higher-quality audits, both from an auditor reputation and a monitoring strength perspective, than do non-brand name auditors.

Information credibility depends upon the auditor's ability to influence the confidence that users place on the information provided in the financial statements. Auditors provide investors with independent assurance that the firm's financial statements conform to GAAP. The fact that stock prices react to earnings announcements suggests that, overall, investors perceive earnings information as credible (Krishnan 2003). Many studies confirm that capital providers require firms to hire an independent auditor as a condition of financing, even when it is not required by the regulations (Ashbaugh and Warfield 2003; Datar et al. 1991). For example, banks normally require firms to present the audited financial information, even for private firms, in making loan and other financial decisions. This implies that capital providers believe auditors can enhance the credibility of financial information provided by firms (Dye 1993; Easton and Pae 2004; Felix et al. 2005).

External auditors are thought to provide value by adding to the credibility of financial reporting (Porter, Simon and Hatherly 2003). Dopuch et al. (1986) suggest that since brand name auditors have more observable characteristics associated with audit quality (e.g., specialized training and rigorous peer reviews), it is expected that stakeholders perceive them as providing greater assurance to the credibility of the financial statements. On the other side, since the credibility of financial statements is judged by users (Dopuch and Siminic 1982), prior research has sought to represent the perceived audit quality through some proxies for the credibility of financial reporting. The common belief shared by bankers and underwriters is that brand name auditors add more credibility to the financial statements (Feltham et al. 1991; Hay and Davis 2004).

4.1.2 Earnings Response Coefficients

Earnings quality is a concept that is not directly observable. The ultimate test of earnings quality is the market's responses to earnings. Capital market-based research has focused on the determination of the earnings response coefficients (ERCs). ERCs is commonly estimated as the slope coefficient in a regression of the abnormal stock returns on a measure of earnings surprise[7]. It is therefore a measure of the extent to which new earnings information is capitalized in the stock price (Kim and Kross 2005; Ryan and Zarowin 2003). Whether and how financial information is used by investors depends on the quality and credibility of the information. Holthausen and Verrecchia (1988) documented a positive association between the magnitude of the stock price responses and the precision of the accounting information.

Teoh & Wong (1993) and Balsam et al. (2003) suggested that investors' responses to an earnings surprise depend on the perceived quality and credibility of the earnings reported. Specifically, Teoh and Wong (1993) hypothesized that to the extent that investors perceived Big 8 auditors as providing higher-quality audits, i.e., as reporting earnings of higher quality and credibility for their clients, the stock price reaction to the unexpected reported earnings of Big 8 clients should be greater than that of others. Consistent with this hypothesis, they found the ERCs of Big 8 clients to be significantly higher than that of non-Big 8 clients. Thus, by linking financial reporting results to the ERCs, they provide evidence that the financial statements audited by Big 8 are of higher quality and utility.

---

[7] ERCs can be estimated from the regression: $AR = \beta_0 + \beta_1 UE + \varepsilon$, where AR is the abnormal stock returns, UE is the unexpected earnings, and β_1 is the estimated value of ERCs.

4.1.3 Earnings Management in the US and in China

Earnings management involves the selection of accounting procedures and estimates that may induce a bias to achieve certain objectives. Earnings management has the potential to decrease the perceived quality of earnings (Schrand and Wong 2003; Watkins et al. 2004). Because earnings management is costly to detect, the agency theory predicts that earnings management will occur when the benefit of manipulating earnings exceeds the relevant costs (Watts and Zimmerman 1986). A large number of studies have documented that business managers have great incentives to manage earnings. These incentives arise out of the explicit and implicit contracts that link the management's interest to accounting numbers (Anderson et al. 2004; Bartov and Mohanram 2004).

The literature on earnings management makes three general predictions about the use of discretion relevant to a particular benchmark (Healy 1985; Abarbanell and Lehavy 2003). First, if pre-managed earnings are high above a relevant benchmark, firms will make income-decreasing choices. Next, if pre-managed earnings are below a benchmark but reserves are available to meet the benchmark, firms will draw from accounting reserves to just beat the benchmark. Finally, if pre-managed earnings and available reserves are insufficient to meet any benchmark, firms will engage in extreme, income-decreasing behavior that will pay back in the future. Healy (1985), Das and Zhang (2003), and Dechow et al. (2003) all reported the evidence from the US market that firms might undertake extreme, income-decreasing earnings management to maximize accounting reserves for future use (i.e., taking an 'earnings bath').

Prior research also documents other incentives for earnings management. Watts and Zimmerman (1990) contended that larger firms would make income-decreasing accounting choices in an attempt to minimize political costs. Other empirical studies (Cahan 1992; Han and Wang 1998; Jones 1991; Cahan et al. 1997; Monem 2003) also document that business managers use income-reducing discretionary accruals to minimize the political costs. Litigation concerns discourage firms to manage earnings upward because of the asymmetric loss function on firms who exaggerate their earnings. Perry and Williams (1994) reported the evidence of downward earnings management in the year prior to the management buyouts in the US.

A large body of recent literature indicates that the management has strong incentives to report earnings growth (e.g., Kothari 2001). Studies by Hayn (1995), Burgstahler and Dichev (1997), and Degeorge et

38

al. (1999) documented unusual patterns in the distribution of earnings levels, earnings changes, and earnings surprises. For instance, Burgstahler and Dichev (1997) reported that an unusually large number of firms report small profits and an unusually small number of firms report small losses. These patterns are widely interpreted as the evidence of earnings management. These findings have spawned many studies to investigate various aspects of the management's incentives to meet or beat earnings benchmarks, including Healy and Wahlen (1999), Dechow and Skinner (2000), Fields et al. (2001), Bartov et al. (2002) and Kasznik and McNichols (2002).

The survey evidence in Graham et al. (2005) indicates that reporting increases in quarterly EPS is an important goal for the management, and may even be more important than either beating analyst forecasts or reporting profits. While Degeorge et al. (1999) provided evidence that the management's first objective was to report positive earnings, then to increase quarterly earnings, and then to beat analyst forecasts. Myers et al. (2005) demonstrated many more firms reported long strings of consecutive increases in EPS than would be expected by chance. They interpreted this phenomenon as an evidence of earnings management and provided the evidence that business managers had incentives to maintain their firms' earnings trends.

Hence, in the US and also in other Western countries, as the management is generally motivated to manipulate earnings upward, there are also bunch of incentives for them to manipulate earnings downward. However, the incentives and behaviors of earnings management are different in China, because the ownership structure and incentive systems of the Chinese listed firms are quite different from their counterparts in the Western world (Haw et al. 2005).

Abarbanell and Lehavy (2003) suggest that firms with poor performance may not take an 'earnings bath' if they want to maintain good investor relations. The Chinese listed firms usually have a highly concentrated ownership structure, which makes the opaqueness gains possible for the controlling owners of the firms. In order to maintain the opaqueness gains, the controlling shareholders may try hard to package the listed firms so as to avoid the monitoring from outside shareholders. They are very unwilling to take an 'earnings bath', since such an action will badly damage the investor relations and very probably lead to much stricter surveillance by market regulators[8]. The Chinese listed firms are willing to take a big bath only if they are labeled as ST firms[9] by the stock exchanges.

As Chinese listed firms are usually controlled by the government or parent state-owned enterprises (SOEs), they face much fewer political costs compared with their US counterparts. The threat of litigation is less likely to deter Chinese business managers from reporting optimistically when it is difficult to successfully sue them for doing so. In China, the legal system is not highly independent, often influenced by the government; and the law enforcement is weak. In addition, class legal action is not allowed in China at present. As the controlling shareholders are usually the government agencies or parent SOEs, it is very difficult for Chinese investors to effectively sue business managers (usually appointed by the controlling shareholders) and the controlling shareholders. Management buyout is nearly impossible as business managers of the Chinese listed firms are usually appointed by the government and they hold none or an insignificant amount of the firm's shares. Kim et al. (2003) found that, in the US, the share-decreasing firms have the intentions to manage earnings downward; however, share-decreasing action is generally disallowed in China.

Considering the facts and the arguments above, the Chinese listed firms have few incentives to manage earnings downward. However, they do have strong incentives to manipulate earnings upward. A significant portion of the Chinese listed firms are owned by the state and legal entities (mostly parent SOEs), which suggests that business managers may have a greater motivation to act in the best interest of the state and parent SOEs in terms of their political performance rather than in the best interest of

---

[8] Actually Chinese listed firms took a market-wide big bath in 2001, as the listed firms are required to make provisions for eight items to follow more conservative accounting treatments. However, this action is initiated by the change of accounting standards but not by the management of listed firms. Our study in this chapter covers years of 2002-2004, and does not include Year 2001.
[9] Special Treatment (ST) stock is designed for a listed company on the Shanghai or Shenzhen stock exchanges when it demonstrates an abnormal financial situation. The ST firms will be delisted if they suffer three consecutive years of loss, so they have the incentive to take big baths to avoid making a loss in the third year. The ST firms are not many (around 5% of all listed firms) and happen to fall outside of our sample.

public shareholders. In order to maintain the opaqueness gains from weak corporate governance mechanism, the controlling owners (through the management) are motivated to manipulate earnings upward to avoid stringent market surveillance and the monitoring from the public investors. For example, the controlling owners have been involved actively in a lot of related-party transactions to boost profit or decorate the below-the-line items as operating income. The listed firms need to manipulate earnings upward to avoid being de-listed or being capped with ST status. As business managers are usually appointed by the government or the controlling SOEs, there are incentives for them to overstate earnings to please their superiors to maximize their own benefits. Business managers are also motivated to manipulate earnings upward to fulfill certain political agenda to get promotion in their political career. In addition, the bonus system also motivates business managers to overstate earnings.

Hence, in China, the listed firms have a much stronger intention to manage earnings upward than to manage earnings downward. We conjecture that the earnings management in China is dominantly unidirectional. Haw et al. (2005) suggested that Chinese investors would differentiate the quality of earnings and put less value on the earnings suspected of a greater degree of manipulation. Their results imply that the Chinese investors, to some extent, can make rational adjustment for the quality of earnings.

4.1.4 Auditors' Role in Curbing Earnings Management

Information of accruals-based earnings is considered superior to cash flows because it overcomes the timing and mismatching problems inherited in the determination of cash flows (Dechow 1994). Accruals allow the management to communicate their private and inside information and thereby improve the ability of earnings to reflect the underlying economic value (Durtschi and Easton 2005; Ghosh et al. 2005). However, business managers could engage in opportunistic earnings management that would seriously undermine the informativeness of the reported earnings (Hribar et al. 2006; Hunton et al. 2006). This phenomenon is especially serious in China, where business managers have strong incentives to opportunistically manipulate earnings upward for the benefits of the controlling owners and themselves.

Investors rely heavily on the audited financial statements and auditor opinions to make investment decisions. If the audited financial statements and auditor opinions turn out to be misleading, investors may sue the auditors. Auditors are thus encouraged to curb opportunistic management of accruals to make earnings credible and of high quality. The audit quality is one factor that restricts the extent to which business managers can manipulate earnings (Balsam et al. 2003; Imhoff 2003; Kadous 2000). A number of studies have examined whether the audit quality, measured by auditor brand name, is associated with the earnings quality. Becker et al. (1998) and Reynolds and Francis (2000) argue that high-quality auditors are more able to detect earnings management because of their superior knowledge and resources. High-quality auditors will curb opportunistic earnings management to protect their reputation. Francis and Krishnan (1999) and Krishnan (2003) reported that Big 6 auditors were more effective in constraining the opportunistic reporting of the discretionary accruals than non-Big 6 auditors.

The literature on auditor characteristics further suggests that auditors play two valuable roles in capital markets: an information role and an assurance role (Dye 1993). Auditors provide independent verification of the financial statements prepared by management, and can detect and report significant breaches in clients' accounting system (Watts and Zimmerman 1981; DeAngelo 1981). As a result, audit quality contributes to the credibility of financial disclosure. In addition, since investors often use the audited financial statements as the basis for asset allocation decisions, the securities laws normally provide recourse protection for investors against the auditors. In this way, auditors provide investors a means to indemnify potential losses (Kellogg 1984; Chow et al. 1988; Stice 1991; Dye 1993). Lennox (1999) found even stronger support to the 'deep pocket hypothesis' than the 'reputation hypothesis'. Therefore, auditors produce value to the capital market. As the high-quality auditors (e.g., Big 4) are more effective in constraining the opportunistic earnings manipulation and provide 'deeper pockets,' they perform better information and assurance roles. Hence, firms audited by large auditors are valued more by investors and have higher ERCs (Teoh and Wong 1993; Beck and Wu 2006; Felix et al. 2005).

4.1.5 Auditor Switch and its Market Implications

Auditors may issue an unclean opinion on the accounting methods and disclosure policies chosen by the management under one or more of the following reasons: limitations on the scope of the auditor's examination; lack of conformity with GAAP; a departure from an accounting principle set by the authorized body; lack of consistency in accounting treatments; division of responsibility between the principal auditor and other auditors; uncertainties in business operations; and emphasis of special issues (SAS No. 58). However, receiving an unclean auditor opinion would depress the price of a firm's securities and impair its ability to raise funds in the future. Therefore, firms receiving qualified auditor opinions may initiate a search for a new auditor whose views are more in line with that of the management. Chow and Rice (1982) found a significant positive association between the qualified opinions and the subsequent auditor switch in the US. In a similar study of Australian firms, Craswell (1988) reported the evidence consistent with Chow and Rice (1982). Krishnan (1994) and Schauer (2002) have also documented similar results.

Auditor switch after a qualified opinion may represent bad news, because the client may be perceived to shop for a clean opinion from a new auditor. Chaney and Philipich (2002) found that stock prices declined after firms' auditor switch. They attribute this phenomenon to the notion that a perceived decline in audit quality impairs the credibility of financial reporting, i.e., it may result in a lower level of assurance to investors and a higher probability that the reported earnings and book values are overstated. However, other studies that examine the information role of audits (proxied by market reactions to auditor switches) find no evidence that changing auditor affects the stock prices (Nichols and Smith 1983; Johnson and Lys 1990; Klock 1994; Rosner 2003). The inconsistent empirical results may suggest that it is necessary to probe into the types of auditor switch to obtain more convincing results.

4.2 Hypothesis Development

The dual characterization of auditors, as both an assurance provider and as an information intermediary, suggests that audits provide value to the capital markets. The financial statements audited by high-quality auditors are perceived to be more credible and hence to reflect true economic value at a greater extent. Considering the audit quality to be commensurate with the size of auditor, we expect Top 10 auditors to provide high-quality audits in China. Furthermore, high-quality auditors have 'deep pockets' and can therefore provide investors with greater assurance in the event of securities litigation (Dye 1993). Mansi et al. (2004) suggested that the assurance effect of audits is value-added to capital market participants and investors value the assurance role of auditors in addition to their information role.

Potential costs associated with audit failure (including a loss of reputation capital) are likely to be higher for Top 10 auditors than for non-Top 10 auditors. For example, when the allegation of audit failure arises, Top 10 auditors are likely to face greater publicity in the financial media, and thus they are likely to bear a greater reputation loss than non-Top 10 auditors. Given the potential litigation risks,

Top 10 are more likely to be sued (and suffer larger damage) because of their perceived "deep pockets;" or, litigation is likely to be more costly for Top 10 auditors in terms of the potential impairment to their brand name reputation capital. Since Top 10 have more to lose in terms of the reputation capital and the size of damage penalty, litigation risks can be expected to motivate large auditors to provide higher-quality audits.

Prior research finds that auditors will curb the management's earnings manipulation and high-quality auditors are more effective in constraining the opportunistic reporting of discretionary accruals (Becker et al. 1998; Francis and Krishnan 1999; Krishnan 2003). Teoh & Wong (1993) and Balsam et al. (2003) found that higher audit quality is associated with lower-level earnings management and higher-level ERCs and that the credibility of financial statements is enhanced. They argue that an earnings surprise will result in a greater stock price reaction when investors perceive the reported earnings to be more credible. Their research is conducted in countries with both directions of earnings management incentives, namely, income-increasing and income-decreasing incentives.

In China, the listed firms have much stronger motivations to manage earnings upward than to manage earnings downward, therefore, the earnings management is conjectured dominantly in the direction of increasing the reported earnings. Correspondingly, the auditors' role in curbing earnings management is conjectured predominantly in the direction of reducing management-reported earnings. Since lawsuits typically allege that earnings were overstated in China, we further argue that Top 10 auditors will be more conservative than non-Top 10 auditors in determining the reported earnings. As a result, Top 10 auditors have a greater intention to limit the management's ability to choose income-increasing accruals than non-Top 10 auditors.

Like previous studies (Teoh and Wong 1993; Becker et al. 1998), we measure firms' earnings quality by the size of ERCs, i.e., the magnitude of the stock market's reaction to unexpected earnings. As Chinese investors, to certain extent, can see through the earnings management activities (Haw et al. 2005), the ERCs of firms audited by Top 10 will be different from those audited by non-Top 10. However, because auditors are perceived mainly in curbing income-increasing earnings manipulation in China, we posit that the impact of high-quality audits on the size of ERCs will be sensitive to the sign of the unexpected earnings, i.e., whether the unexpected earnings is positive or negative.

44

For the same magnitude of positive unexpected earnings, the firms audited by Top 10 are perceived to be more conservative and report less optimistically. Responding to the firms' conservatism, investors will appreciate their stock prices more, thus, implying higher ERCs for the firms audited by Top 10. For the same magnitude of negative unexpected earnings, firms audited by Top 10 are perceived to be more conservative and report more pessimistically. Similarly, responding to the firms' conservatism, investors will depreciate their stock prices less, implying lower ERCs for the firms audited by Top 10. To examine the effect of the audit quality on ERCs, we test empirically the following hypotheses:

Hypothesis 7a (H7a): ceteris paribus, for positive unexpected earnings, firms audited by Top 10 auditors have higher ERCs than those audited by non-Top 10 auditors.

Hypothesis 7b (H7b): ceteris paribus, for negative unexpected earnings, firms audited by Top 10 auditors have lower ERCs than those audited by non-Top 10 auditors.

Prior studies argued that the market reaction to auditor switch is negative because the switch signifies that the firm may be attempting to influence the auditor by shopping for audit services (Chow and Rice 1982). However, previous empirical studies report inconsistent results. Chaney and Philipich (2002) found that stock prices declined after firms switched their auditors. They attribute this phenomenon to the notion that a perceived decline in the audit quality impairs the credibility of financial reporting, which may result in a lower level of assurance to investors and a higher probability that the reported earnings and book values are overstated. Other studies that examine the information role of audits (proxied by market reaction to auditor switch) found little evidence that changing auditor would affect the stock prices (Nichols and Smith 1983; Johnson and Lys 1990; Klock 1994; Schauer 2002). We conjecture that the inconsistent results arise because the prior studies did not differentiate firms switching to lower-quality auditors from those switching to higher-quality auditors.

Information asymmetry leads to a demand for effective corporate governance mechanisms, which can be used to enhance the credibility of financial information. The choice of a high-quality auditor is one possibility. The audit quality can be used as a signal of auditor's value (information signaling) and assurance provided (assurance signaling) (Willenborg 1999). Similarly, auditor switch also signals the inside information to investors. When firms switch auditors, they can switch to a higher-quality auditor or a lower-quality auditor. If the successor auditor is of higher quality to the predecessor auditor, the signal is that the management may not be able to make aggressive income-increasing earnings manipulation. On the contrary, if the successor auditor is of lower quality than the predecessor auditor,

45

it may indicate that the management wants to manipulate earnings aggressively and prefers a more pliable auditor. Assuming the market is efficient and investors can differentiate earnings quality, we expect that, for switching to higher-quality auditors, the market will attach higher ERCs to firms with good news and lower ERCs to firms with bad news; and for switching to lower-quality auditors, the market will attach higher ERCs to firms with bad news and lower ERCs to firms with good news. To examine the effect of auditor switch on ERCs, the following hypotheses are tested empirically:

Hypothesis 8 (H8): ceteris paribus, for positive (negative) unexpected earnings, firms switching to larger auditors have higher (lower) ERCs than other firms.

Hypothesis 9 (H9): ceteris paribus, for positive (negative) unexpected earnings, firms switching to smaller auditors have lower (higher) ERCs than other firms.

4.3 Research Methodology

4.3.1 Model Specification

Prior studies have examined the market implications of auditor switch, but they did not differentiate the types of auditor switch. Our study intends to empirically investigate this issue that has not been explored in the extant literature. To test Hypotheses 7 to 9, we run Equation (3). As outlined in the previous section, auditor quality and auditor switch will influence ERCs oppositely in terms of the opposite signs of abnormal earnings, so we divide all firm-year observations according to the sign of the abnormal earnings and test the four hypotheses separately. As the role of financial analysts is very limited in the Chinese stock market, we apply the random walk model to perform the tests. In the random walk model, earnings of last year is assumed to be the expected earnings, therefore the abnormal earnings is the difference between earnings of this year and that of last year. To eliminate the size effect, the abnormal earnings is usually deflated by the market value of equity. Hence, $\Delta E = (E_t - E_{t-1}) / MV$, where ΔE is the abnormal earnings, E_t is earnings of this year, E_{t-1} is earnings of last year, and MV is the market value of equity.

The interaction terms are used to test H7 to H9. Thus, T10*ΔE (Top 10 auditor * abnormal earnings), SU*ΔE (switching to a larger auditor * abnormal earnings) and SD*ΔE (switching to a smaller auditor * abnormal earnings) are used to test H7, H8 and H9 respectively. However, we do not incorporate

46

some independent variables, namely T10 (Top 10 auditor), SU (switching to a larger auditor) and SD (switching to a smaller auditor) in Model (3). The reason is that we adopt the two-day short window (the earnings announcement day and the immediately following day) in the test. For the short-window study, T10, SU and SD are all old information and should have been reflected in stock prices before the earnings announcement date. Hence, these independent variables (T10, SU and SD) should not affect the two-day cumulative abnormal return (CAR). In the study we use the market-adjusted model to calculate CAR: $CAR_i = (r_{i0}-r_{m0}) + (r_{i1}-r_{m1})$, where r_{i0} and r_{i1} are the stock return for firm i on the annual earnings announcement day and on the immediately following day respectively, and r_{m0} and r_{m1} are the market return on the annual earnings announcement day and on the immediately following day respectively. We expect δ_2 and δ_3 to be positive (negative) and δ_4 to be negative (positive) for firms with positive (negative) abnormal earnings.

$$CAR = \delta_0 + \delta_1\Delta E + \delta_2 T10*\Delta E + \delta_3 SU*\Delta E + \delta_4 SD*\Delta E + \delta_5 OPI + \delta_6 OPI*\Delta E +$$
$$\delta_7 MB*\Delta E + \delta_8 ABBETA*\Delta E + \varepsilon \qquad\qquad (3)$$

Where:

CAR	=market-adjusted abnormal returns accumulated over the two trading days (0, +1), where 0 is the annual earnings announcement date, adjusted for dividends and stock rights
ΔE	=unexpected annual earnings, calculated as earnings of this year minus earnings of last year and then divided by market value of equity
T10	=1 if the firm is audited by a Top 10 auditor, and 0 otherwise
SU	=1 if the firm switches to a larger auditor in the current or the last year, and 0 otherwise
SD	=1 if the firm switches to a smaller auditor in the current or the last year, and 0 otherwise
OPI	=1 if the firm receives an unclean auditor's opinion in the current year, and 0 otherwise

MB	=market-to-book ratio at the end of the previous year, calculated as the market value of the firm's stock divided by its book value
ABBETA	=the absolute value of beta to proxy for firm risk

Model (3) controls for several other factors, including the auditor's opinion, firm growth and risk. Dopuch et al. (1986) found that the qualified auditor opinions were associated with negative stock price reactions. Chen et al. (2000) reported the evidence that there is a negative association between the modified auditor opinions (MAOs) and market reactions in the Shanghai Stock Exchange. Ghosh et al. (2005) and Ertimur et al. (2003) both found that the earnings response coefficients (ERCs) of growing firms were significantly larger than those of other non-growing firms, indicating that investors attach a relatively greater importance to the earnings of the growing firms. In addition, the dividend discount model assumes that stock price and risk are negatively correlated, since a higher discount (risk) factor will result in a lower stock price, holding the expected dividends constant. Early empirical research on ERCs such as Collins and Kothari (1989) and Easton and Zmijewski (1989) also reported that ERCs were negatively related to firms' risk. Hence, we expect δ_5, δ_6 and δ_8 to be negative and δ_7 to be positive for firms with positive unexpected earnings, and δ_5 to be negative and δ_6 to be positive for firms with negative unexpected earnings. Because of the distinctive nature of negative unexpected earnings, we do not predict the signs of δ_7 and δ_8 for firms with negative unexpected earnings.

4.3.2 Sampling

Our sample consists of the listed firms with annual earnings announcements during the fiscal years 2002-2004. Unlike the prior two tests reported in Chapter Two and Chapter Three respectively, we do not include 2001 observations in the market response tests. This is because that we will use year t-1 information for SD (switching to a smaller auditor) and SU (switching to a larger auditor) classification in the market response tests. In a preliminary test, we found that SD and SU in year t-1 have a significant effect on two-day CAR on the abnormal earnings in year t. Therefore, to expand our sample size, we use the SD and SU information in both year t-1 and year t in the tests. Data are collected from CSMAR Trading Database, CSMAR Financial Database, TEJ, China Security Daily, Shenzhen Security Times, and Shanghai Security News. The finance, transportation, and utility firms are excluded because of their distinctive natures. We also exclude firms that change CEOs in the period

2002-2004 because those firms may take big earnings baths and thus may contaminate the empirical results. Furthermore, we exclude firms who switch auditors for more than once during the test period because those firms may conduct two types of auditor switch and they are out of our research scope. Descriptions of the data for firms with positive and negative abnormal earnings are provided in Table 6 and Table 7 respectively.

[insert Table 6 here]

[insert Table 7 here]

There are 3,587 firm-year observations for A-share listed firms during 2002-2004, among which 1,961 observations (representing 54.7% of the total) are with positive abnormal earnings and 1,626 observations (representing 45.3% of the total) with negative abnormal earnings.

Panels A of Table 6 and Table 7 derive the final sample size of 1,284 and 981 for firms with positive and negative abnormal earnings respectively, after deleting observations with missing values, observations for the financial, transportation, and utility sectors, observations with CEO turnovers, and observations with auditor switch more than once. Panels B of Table 6 and Table 7 present the descriptive statistics of the variables. The CSRC sets the daily price fluctuation limit to be 10% for the listed firms in China, so CAR for any two days are always between -20% to +20% (with rounding errors). Unlike in the US, where large auditors (Big 4) apparently dominate the audit market, Top 10's market share is much smaller in China. Top 10 have audited 23.4% (301/1,284) of firms with positive abnormal earnings and 21.1% (207/981) of firms with negative abnormal earnings. As for auditor's opinion, 4.0% of firms with positive ΔE and 10.7% of firms with negative ΔE respectively, received unclean auditor's opinions during the three-year period. Panels C of Table 6 and Table 7 present the correlation coefficient matrix. For observations with positive abnormal earnings, the two-day CAR is positively related with SU (switching to a larger auditor) and negatively related with ABBETA (the absolute value of beta) at the significant level. Correlation coefficients among independent variables are not high, with the largest one at the level of 0.273. For observations with negative abnormal earnings, the two-day CAR is positively related with ΔE (abnormal earnings), T10 (Top 10 auditor), and SU (switching to a larger auditor) and negatively related with SD (switching to a smaller auditor), OPI (unclean auditor opinion) and ABBETA (the absolute value of beta to proxy for firm risk) at the significance level. Correlation coefficients among the independent variables are also not high with the

49

largest one at the level of -0.465 (between the abnormal earnings and an unclean auditor opinion for observations with negative abnormal earnings).

4.4 Empirical Results

4.4.1 Results for Firms with Positive Abnormal Earnings

Table 8 provides empirical regression results for firms with positive abnormal earnings. F-statistic is 7.287, at 0.000 significance level, suggesting that the whole model is significant in predicting the abnormal stock returns. The adjusted R-square is 3.8%, similar to the level reported by prior studies. The coefficient for ΔE is positively significant (δ_1=0.083, t=1.657, and significant at 10% level), suggesting that accounting earnings is informative for investors. Though the Chinese stock market is weakly regulated with lots of information leakage, investors still use accounting information in making investment decisions.

[insert Table 8 here]

Consistent with H7a, δ_2 is positively significant (δ_2=0.198, t=3.817, and significant at 1% level), suggesting that firms audited by Top 10 auditors have higher ERCs. In other words, stock prices of Top 10 clients have a higher price appreciation for good news. This can be interpreted as the Chinese investors may have perceived that there are lower income-increasing accruals in Top 10 clients' reported earnings compared with that of non-Top 10 clients. Consistent with H8, δ_3 is positively significant (δ_3=0.266, t=4.588, and significant at 1% level), suggesting that firms switching to a larger auditor have higher ERCs than other firms. Investors may perceive switching to a larger auditor as a signal of high earnings quality and give the stock price a bigger appreciation. H9 is also supported, with δ_4 significantly negative (δ_4=-0.215, t=-3.108, and significant at 1% level). It suggests that investors appreciate the stocks of firms with good news to a less extent if the firms switch to a smaller auditor.

The coefficient for MB*ΔE (market-to-book ratio * abnormal earnings) is significant (δ_7=0.002, t=1.729, and significant at 10% level), suggesting that the Chinese investors put more value on the growing firms. Consistent with prior studies (Collins and Kothari 1989; Easton and Zmijewski 1989), the coefficient for ABBETA*ΔE (absolute value of beta * abnormal earnings) is negative (δ_8=-0.138,

50

t=-3.551, and significant at 1% level), implying lower ERCs for riskier firms. Coefficients for OPI (unclean auditor opinion) and OPI*ΔE are insignificant, and the reason may be that we did not separate the different types of unclean auditor opinions (i.e., unqualified with explanation, qualified, adverse, and disclaimer).

In summary, the empirical results support H7a, H8 and H9. Firms audited by Top 10 auditors are perceived to have less income-increasing earnings management and therefore investors believe their financial information are more credible, leading to higher ERCs to the unexpected earnings of these firms. Switching to a larger (smaller) auditor signals high (low) earnings quality, and thus increases (decreases) ERCs for the firms. These findings may suggest that Chinese investors are relatively sophisticated in using accounting information to make investment decisions.

4.4.2 Results for Firms with Negative Abnormal Earnings

Table 9 presents empirical results for firms with negative abnormal earnings. For this regression, F-statistic is 7.712, significant at 0.000 level, suggesting that the whole model is significant in predicting the abnormal stock returns. The adjusted R-square is 5.2%, a similar level to that of prior studies. The coefficient for ΔE is positive but insignificant (δ_1=0.004, t=0.069). This result is not surprising as financial reporting normally modifies the revenue recognition rules by adopting a lower verification standard for information about the decreases in earnings than for the increases. This type of conservatism is an asymmetric response to uncertainty, thus leading to an asymmetric response of stock returns to the abnormal earnings. For instance, Basu (1997) found that ERCs were higher for positive earnings changes than for negative earnings changes. This result suggests that the Chinese investors will evaluate good news and bad news differently.

The coefficient for T10*ΔE (Top 10 auditor * abnormal earnings) is negative and significant (δ_2=-0.148, t=-3.885, and significant at 1% level), supporting H7b. The result suggests that as firms audited by Top 10 are perceived more conservative in earnings reporting, their stocks receive smaller price depreciation at the announcement of bad news. Consistent with H8 and H9, coefficients for SU*ΔE (switching to a larger auditor * abnormal earnings) and SD*ΔE (switching to a smaller auditor * abnormal earnings) are negative (β_3=-0.222, t=-4.064, and significant at 1% level) and positive (β_4=0.106, t=2.308, and significant at 5% level) respectively, implying that the auditor switch has a signaling effect to the market. Switching to a larger auditor signals that earnings is conservative and of

51

high quality, therefore firms will experience a smaller decline of stock prices upon bad news. Vice versa, switching to a smaller auditor signals that earnings is probably less conservative and the bad news may actually even be worse than reported. Hence, firms do face a bigger depreciation of stock prices.

Consistent with the predictions, we also find that δ_6 is positively significant (δ_6=0.083, t=1.958, and significant at 10% level), implying that receiving an unclean auditor's opinion will further depreciate stock prices. In summary, H7b, H8 and H9 are supported. While firms audited by Top 10 receive bigger stock price appreciation for good news, they experience smaller stock price depreciation for bad news. Firms switching to a larger auditor are awarded with smaller stock price declines at bad news, while those switching to a smaller auditor are punished with bigger stock price depreciations. These findings further testify that the Chinese investors' are relatively sophisticated in making investment decisions.

4.5 Sensitivity Tests

We pooled the data of three years together in the main tests. To examine the robustness of the empirical results, we further run the regression by yearly data. Table 10-1, 10-2 and 10-3 presents the results of robustness tests for firms with positive abnormal earnings. The regression results are similar to the main tests. The coefficients for T10*ΔE (Top 10 auditor * abnormal earnings) and SU*ΔE (switching to a larger auditor * abnormal earnings) are positively significant for all three years, and those for SD*ΔE (switching to a smaller auditor * abnormal earnings) are negatively significant for all three years in a consistent way, implying that our empirical results are robust for tests of positive abnormal earnings. Table 11-1, 11-2 and 11-3 provides the robustness test results for firms with negative abnormal earnings. The coefficients for T10*ΔE (Top 10 auditor * abnormal earnings) and SU*ΔE (switching to a larger auditor * abnormal earnings) are negatively significant for all three years, and those for SD*ΔE (switching to a smaller auditor * abnormal earnings) are positively significant for all three years. Therefore, H7b, H8 and H9 are also well supported.

[insert Table 10-1 here]

[insert Table 10-2 here]

[insert Table 10-3 here]

[insert Table 11-1 here]

[insert Table 11-2 here]

[insert Table 11-3 here]

The empirical results remain substantially the same after adopting the following changes in the regression tests respectively: using three-day abnormal returns; deleting the observations with three standard deviations away from the mean. In sum, the empirical results of our study are robust to the sensitivity tests.

Chapter Five

Conclusions

5.1 Summary

The purpose of this study is to investigate the association between firms' internal corporate governance mechanism and their auditor choice and auditor switch decisions and how investors respond to firms' auditor choice and auditor switch decisions in the Chinese context. Three measures are used to proxy for firms' internal corporate governance mechanism, i.e., the ownership concentration (shareholding of the controlling owner), the size of SB, and the duality of the CEO and the BoD chairman positions. For auditor choice tests, we categorize all auditors into two groups, Top 10 and non-Top 10 according to the ranking of the Chinese Institute of Certified Public Accountants (CICPA) in terms of the annual revenue of auditing firms. For auditor switch analysis, we divide all auditor switches into two types: switching to a larger auditor and switching to a smaller auditor. We use ERCs to test the investors' responses to firms' auditor choice and auditor switch in respect of the firms' stock price changes responding to their abnormal annual earnings.

Three sets of hypotheses are developed. Firstly, H1 to H3 are used to test the association between the firms' internal corporate governance mechanism and auditor choice decisions. The empirical results support the three hypotheses. Firms with larger controlling owners, with smaller SB size, and in which the positions of the CEO and the BoD chairman are held by the same person are less likely to hire a Top 10 auditor. The listed firms in China are highly concentrated in ownership and are controlled by the largest owners (usually the government or parent SOEs). For controlling owners in the Chinese listed firms, the sole dominant purpose of getting firms listed is to raise capital. The controlling owners may try to adopt effective external monitoring mechanisms (such as high-quality audits), but their main goal is to improve company image for the benefit of raising capital in the future. When new equity issuing is unlikely, the controlling owners may be less enthusiastic in improving corporate governance of their firms. Capitalization of the opaqueness gains from weak corporate governance mechanism will then dominate firms' auditor choice decisions. The listed firms are inclined to choose auditors of low quality to maintain their opaqueness gains. With more opaqueness gains to protect, firms with weaker internal corporate governance mechanism are more likely to choose a low-quality auditor.

H4 to H6 are used to test the association between firms' internal corporate governance mechanism and auditor switch decisions. H4 and H6 are supported, while there is insufficient evidence for H5. Firms with larger controlling owners and in which the CEO and the BoD chairman are the same person are more likely to switch to a smaller auditor. However, whether firms with smaller SB opt to switch to a smaller auditor is in doubt. Generally, we conclude that firms with weaker internal corporate governance mechanism are more likely to switch to a smaller auditor to sustain their opaqueness gains. The coefficient for the size of SB is not significant at the conventional level. This may imply that SB does not perform an effective monitoring role, because the members of SB are mainly from inside the firms.

For the tests of market implications of auditor choice and auditor switch decisions, H7 to H9 are all well supported. Firms with positive (negative) abnormal earnings audited by Top 10 auditors or switching to a larger auditor have higher (lower) ERCs to unexpected annual earnings. Firms with positive (negative) abnormal earnings switching to a smaller auditor have lower (higher) ERCs. Following the argument that the auditor size is a good proxy for audit quality, the results may suggest that the Chinese investors can differentiate earnings quality. The fact that the coefficients for auditor switch are significant would further confirm the Chinese investors' sophistication in making investment decisions.

Below is a summary of all hypotheses tested in our study and whether or not they are supported by the empirical evidence.

A Summary of All Hypotheses Tested

Hypotheses	Empirical results
H1: ceteris paribus, the higher percentage of total shares held by the largest owner, the less likely a Top 10 auditor will be chosen	Supported
H2: ceteris paribus, a firm with fewer SB members is less likely to choose a Top 10 auditor	Supported
H3: ceteris paribus, a firm with no duality of the positions for CEO and BoD chairman is less likely to choose a Top 10 auditor	Supported

H4: ceteris paribus, the higher percentage of total shares held by the controlling owner, the more likely the firm will switch to a smaller auditor	Supported
H5: ceteris paribus, a firm with a smaller SB size is more likely to switch to a smaller auditor	Not supported
H6: ceteris paribus, a firm with the CEO and the BoD chairman held by the same person is more likely to switch to a smaller auditor	Supported
H7a: ceteris paribus, for positive unexpected earnings, firms audited by Top 10 auditors have higher ERCs than those audited by non-Top 10 auditors	Supported
H7b: ceteris paribus, for negative unexpected earnings, firms audited by Top 10 auditors have lower ERCs than those audited by non-Top 10 auditors	Supported
H8: ceteris paribus, for positive (negative) unexpected earnings, firms switching to larger auditors have higher (lower) ERCs than other firms	Supported
H9: ceteris paribus, for positive (negative) unexpected earnings, firms switching to smaller auditors have lower (higher) ERCs than other firms	Supported

The findings of this study have theoretical and practical implications. This study expands the literature on auditor choice, auditor switch and market responses to audit quality and auditor switch from the developed markets to the less developed Chinese market. Prior studies (Fan and Wong 2005; Hay and Davis 2004) find out that firms with high agency costs are inclined to choose a high-quality auditor to improve their corporate governance. Different from them, this study reveals that, under certain circumstances, in order to sustain opaqueness gains, firms with weak internal corporate governance mechanism and therefore high agency costs are inclined to avoid high-quality auditors.

Prior studies (Teoh and Wong 1993; Balsam et al. 2003) document that clients of high-quality auditors have higher ERCs than those of low-quality auditors. However, this study suggests that the association between ERCs and audit quality is sensitive to the sign of abnormal earnings. This study further suggests that investors respond differently to different types of auditor switch (i.e., switching to a larger auditor and switching to a smaller auditor).

The study findings may also generate implications for stakeholders of the Chinese listed firms, especially the investors and market regulators. It might be a very bad signal for listed firms with weak internal corporate governance mechanism to choose or switch to a low-quality auditor. In so doing, the controlling shareholders of the listed firms are easy to realize private benefits from exploiting the small shareholders. To bolster the confidence of the market participants, the regulators should carefully monitor auditing practices to protect the interests of investors. The findings also imply that the Chinese investors are relatively sophisticated in making investment decisions. Whether the financial information is of high quality and of high credibility influences the investors' decisions.

5.2 Limitations

There are some limitations in this study. First, for the auditor choice and auditor switch tests, more rigorous results may be derived from more sophisticated simultaneous equation methods. Prior studies (Feltham 1991; Copley and Douthett 2002) suggested that auditor choice research should consider both the demand and the supply sides of audit services. Simultaneous equation methods could be used to control for both the demand and the supply sides of audit services. However, as many Chinese listed firms do not report the audit fee information, controlling for the supply side effects is difficult to realize. Second, the size of SB may not be a good proxy for internal corporate governance mechanism, since SB members are mainly from inside the firms. The size or expertise of independent (non-executive) directors might be a better proxy, but information about independent directors for the testing period is not available from the databases. We also used other variables to proxy for internal corporate governance, such as number of directors, and age and educational background of directors. However, the statistical results are only weakly significant. Third, this study considers only the contemporaneous association between the stock returns and the abnormal accounting earnings. If the study incorporates the price-leading-earnings variables, the findings may be more complete and robust, and the goodness-of-fit may also be improved significantly.

5.3 Future Research Opportunities

In prior studies, the auditor size is argued to be commensurate with the audit quality. We also employ the size of auditor to proxy for audit quality in China. However, due to the distinct characteristics of Chinese audit market, there might be other ways to measure the audit quality. Before 1997, Chinese auditing firms were affiliated to their sponsoring bodies, mainly the government agencies and other social institutions. A program to de-link auditing firms from their sponsoring bodies was introduced by the government in 1997. The objective of the de-linking program is to make auditing firms financially and operationally independent. It is therefore interesting to study whether Chinese auditors have improved their quality after the de-linking program and whether the original affiliation affects audit quality. In addition, other methods may be adopted to test whether auditors of bigger size can provide higher-quality audits in China, such as examining the association between the discretionary accruals of the listed firms and the size of auditors.

In the market implications study, we adopt a short test window (announcement day t_0, and t_{+1}). As there is significant information leakage in the Chinese stock market, it may be meaningful to use a longer window in future studies. The advantage of using a long window is that it may capture stock price changes based on both the information leakage before the public disclosure and the information drift after the disclosure.

We choose the bear market of 2001-2004 to study the impact of the internal corporate governance mechanism on auditor choice and auditor switch decisions. It is found that when the opportunity of equity offering is low, firms with weaker internal corporate governance mechanism are likely to choose small auditors and switch to smaller auditors. Future research may further test whether firms hiring large auditors are more likely to offer equity rights, controlling for other factors.

Appendix: Ranking of Auditors in China

Ranking	Auditor	Ranking	Auditor
1	PwC Zhongtian	44	Jiangsu Tianhua
2	KPMG Huazhen	45	Huayan
3	Deloitte Huayong	46	Gansu Wulian
4	EY Huaming	47	Zhejiang Wanbang
5	Lixin Changjiang	48	Beijing Zhongzhou
6	Yuehua	49	Huazheng
7	Xinyongzhonghe	50	Guangdong Hengxin
8	Beijing Jingdu	51	Nanjing Yonghua
9	Jiangsu Gongzheng	52	Shandong Wanlu
10	EY Dahua	53	Chongqing Tianjian
11	Zhongshen	54	Xi'an Sigma
12	Zhongruihua	55	Jiangsu Tianye
13	Tianzhizixin	56	Anhui Huapu
14	Shanghai Zhonghua	57	Xiamen Tianjian
15	Lianda	58	Sichuan Huaxin
16	Zhejiang Tianjian	59	Shandong Huide
17	Tianjian	60	Beijing Zhongxingyu
18	Guangzhou Yangcheng	61	Beijing Zhongwei
19	Zhongtianhuazheng	62	Shenzhen Tianjian
20	Shenzhen Pengcheng	63	Fujian Huaxing
21	Shanghai Donghua	64	Beijing Zhongxing
22	Tianyi	65	Shanghai Wanlong
23	Hubei Daxin	66	Shandong Tianghengxin
24	Shanghai Gongxin	67	Yatai Group
25	Jiangsu Suya	68	Zhongtianyin
26	Zhongxi	69	Huajian
27	Zhongxingcai	70	Guangdong Kangyuan
28	Wuhan Zhonghuan	71	Shanghai Tongcheng
29	Zhejiang Dongfang	72	Zhonghengxin
30	Beijing Zhongluhua	73	Beijing Zhongzheng
31	Shanghai Shangkuai	74	Shanghai Shangshen
32	Tianjin Wuzhou	75	Liaoning Tianjian
33	Shandong Zhengyuan	76	Beijing Xinghua
34	Shenzhen Nanfang	77	Beijing Zhongtianheng
35	Shenzhen Dahua	78	Shandong Qianju
36	Guangdong Zhengzhong	79	Sichuan Hongri
37	Hu'nan Kaiyuan	80	Zhongqin Wanxin
38	Beijing Yongtuo	81	Hebei Hua'
39	Yunnan Yatai	82	Beijing Zhongpingjian
40	Zhonglei	83	Sichuan Junhe
41	Jiangsu Tianheng	84	Shanghai Jiahua
42	Guangdong Tianhua	85	Guangxi Xianghao
43	Beijing Tianhua		

Notes:

1) The rankings are based on the average audit revenues of Year 2002-04.

2) Auditors must be ranked among the top 100 based on their revenues for all three years
   of 2002-04.

Reference

Abarbanell, J., Lehavy, R., 2003. Can Stock Recommendations Predict Earnings Management and Analysts' Earnings Forecast Error? *Journal of Accounting Research* 41: 1-31.

Abdel-Khalik, A.R., 2002. Reforming Corporate Governance Post Enron: Shareholders' Board of Trustees and the Auditor. *Journal of Accounting and Public Policy* 21(2): 97-126.

Aharony, J., Lee, C.W.J., and Wong, T.J., 2000. Financial Packaging of IPO Firms in China. *Journal of Accounting Research* 38(1): 103-126.

Anderson, U., Kadous, K., and Koonce, L., 2004. The Role of Incentives to Manage Earnings and Quantification in Auditors' Evaluations of Management-Provided Information. *Auditing: A Journal of Practice & Theory* 23(1): 11-27.

Ashbaugh, H., and Warfield T.D., 2003. Audits as a Corporate Governance Mechanism: Evidence from the German Market. *Journal of International Accounting Research* 2: 1-21.

Ball, R., and Shivakumar, L., 2005. Earnings Quality in UK Private Firms: Comparative Loss Recognition Timeliness. *Journal of Accounting and Economics* 39: 83-128.

Balsam, S., Krishnan, J., and Yang, J.S., 2003. Auditor Industry Specialization and Earnings Quality. *Auditing: A Journal of Practice & Theory* 22(2): 71-97.

Bartov, E., Givoly, D., and Hayn, C., 2002. The Rewards to Meeting or Beating Earnings Expectations. *Journal of Accounting and Economics* 33: 173-204.

Bartov, E., and Mohanram, P., 2004. Private Information, Earnings Manipulations, and Executive Stock-Option Exercises. *The Accounting Review* 79(4): 889-920.

Basu, S., 1997. The Conservatism Principle and Asymmetric Timeliness of Earnings. *Journal of Accounting & Economics* 24: 3-37.

Beasley, M., Carcello, J., Hermanson, D., and Lapides, P.D., 2000. Fraudulent Financial Reporting: Consideration of Industry Traits and Corporate Governance Mechanisms. *Accounting Horizons* 14: 441-454.

Beattie, V., and Fearnley, S., 1995. The Importance of Audit Firm Characteristics and the Drivers of Auditor Change in UK Listed Companies. *Accounting and Business Research* 25(100): 227-239.

Beatty, R.P., 1989. Auditor Reputation and the Pricing of Initial Public Offerings. *The Accounting Review* 64(4): 693-709.

Beck, P.J., and Wu, M.G.H., 2006. Learning by Doing and Audit Quality. *Contemporary Accounting Research* 23(1): 1-36.

Becker, C.L., DeFond, M.L., Jiambalvo, J., and Subramanyam, K.R., 1998. The Effect of Audit Quality on Earnings Management. *Contemporary Accounting Research* 15: 1-24.

Bedard, J.C., and Johnstone, K.M., 2004. Earnings Manipulation Risk, Corporate Governance Risk, and Auditors' Planning and Pricing Decisions. *The Accounting Review* 79(2): 277-304.

Berger, P.G., 2003. Discussion of "Differential Market Reaction to Revenue and Expense Surprises". *Review of Accounting Studies* 8: 213-220.

Bloomfield, R.J., 2004. Discussion of "Examining the Role of Auditor Quality and Retained Ownership in IPO Markets: Experimental Evidence". *Contemporary Accounting Research* 21(1): 131-137.

Botosan, C.A., and Stanford, M., 2005. Managers' Motives to Withhold Segment Disclosures and the Effect of SFAS No. 131 on Analysts' Information Environment. *The Accounting Review* 80(3): 751-771.

Burgstahler, D.C., and Dichev, I., 1997. Earnings Management to Avoid Earnings Decreases and Losses. *Journal of Accounting and Economics* 24: 99-126.

Burgstahler, D.C., and Eames, M.J., 2003. Earnings Management to Avoid Losses and Earnings Decreases: Are Analysts Fooled? *Contemporary Accounting Research* 20(2): 253-294.

Cahan, S.F., 1992. The Effects of Antitrust Investigations on Discretionary Accruals: A Refined Test of the Political Cost Hypothesis. *The Accounting Review* 67(1): 77-95.

Cahan, S.F., Charis, B.M., and Elmendorf, R.G., 1997. Earnings Management of Chemical Firms in Response to Political Costs from Environmental Legislation. *Journal of Accounting, Auditing and Finance* 12(1): 37-65.

Canning, M., and Gwilliam, D., 1999. Non-Audit Services and Auditor Independence: Some Evidence from Ireland. *The European Accounting Review* 8(3): 401-419.

Carcello, J.V., Hermanson, D.R., and Neal, T.L., and Riley, R.A., 2002. Board Characteristics and Audit Fees. *Contemporary Accounting Research* 19(3): 365-384.

Carcello, J.V., and Neal, T.L., 2000. Audit Committee Composition and Auditor Reporting. *The Accounting Review* 75: 453-468.

Carcello, J.V., and Neal, T.L., 2003. Audit Committee Characteristics and Auditor Dismissals following "New" Going-Concern Reports. *The Accounting Review* 78(1): 95-117.

Chan, D.K., and Wong, K.P., 2002. Scope of Auditors' Liability, Auditor Quality, and Capital Investment. *Review of Accounting Studies* 7(1): 97-122.

Chan, K.H., Lin, K.Z., and Mo, P.L.L., 2006. A Political-economic Analysis of Auditor Reporting and Auditor Switches. *Review of Accounting Studies* 11(1): 21-48

Chaney, P.K., Jeter, D.C., and Shivakumar, L., 2004. Self-selection of Auditors and Audit Pricing in Private Firms. *The Accounting Review* 79(1): 51-72.

Chaney, P.K., and Philipich, K., 2002. Shredded Reputation: the Cost of Audit Failure. *Journal of Accounting Research* 40(Fall): 1221-1245.

Chau, G., and Leung, P., 2006. The Impact of Board Composition and Family Ownership on Audit Committee Formation: Evidence from Hong Kong. *Journal of International Accounting Auditing & Taxation* 15(1): 1-24.

Chen, C.J.P., Su, X., and Zhao, R. 2000. An Emerging Market's Reaction to Initial Modified Audit Opinions: Evidence from the Shanghai Stock Exchange. *Contemporary Accounting Research* 17(Fall): 429-455.

Chen, C.J.P., Chen, S., and Su, X., 2001. Profitability Regulation, Earnings Management, and Modified Audit Opinions: Evidence from China. *Auditing: A Journal of Practice & Theory* 20(2): 9-30.

Chen, J.J., 2005. China's Institutional Environment and Corporate Governance. *Corporate Governance: A Global Perspective Advances in Financial Economics* 11: 75-93.

Chen, K.C.W., and Yuan, H., 2004. Earnings Management and Capital Resources Allocation: Evidence from China's Accounting-Based Regulation of Rights Issues. *The Accounting Review* 79(3): 645-665.

China Securities Daily. May 16, 1995

.

Chow, C.W., 1982. The Demand for External Auditing: Size, Debt and Ownership Influences. *The Accounting Review* 57(2): 272-291.

Chow, C.W., and Rice, S.J., 1982. Qualified Audit Opinions and Auditor Switch. *The Accounting Review* 57: 326-335.

Chow, C.W., Kramer, L., and Wallace, W., 1988. The Environment of Auditing, in *Research Opportunities in Auditing: the Second Decade* (A.R. Abdel-Khalik and I. Solomon, editors). Sarasota, FL: American Accounting Association.

Citron, D.B., and Manalis, G., 2001. The International Firms as New Entrants to the Statutory Audit Market: An Empirical Analysis of Auditor Selection in Greece, 1993 to 1997. *European Accounting Review* 10(3): 439-462.

Claessens, S., and Fan, J.P.H., 2003. Corporate Governance in Asia: A Survey. *Working Paper.*

Claessens, S., Djankov, S., Fan, J.P.H., and Lang, L.H.P., 2002. Disentangling the Incentive and Entrenchment Effects of Large Shareholdings. *Journal of Finance* 57: 2741-2771.

Cohen, J., Krishnamoorthy, S., and Wright, A.M., 2002. Corporate Governance and the Audit Process. *Contemporary Accounting Research* 19(4): 573-594.

Coles, J.L., Hertzel, M., and Kalpathy, S., 2006. Earnings Management around Employee Stock Option Reissues. *Journal of Accounting and Economics* 41(1/2): 173-200.

Collins, D.W., and Kothari, S.P., 1989. An Analysis of Inter-temporal and Cross-sectional Determinants of Earnings Response Coefficients. *Journal of Accounting and Economics* 11(2/3): 143-181.

Copley, P.A., and Douthett, E.B., 2002. The Association between Auditor Choice, Ownership Retained, and Earnings Disclosure by Firms Making Initial Public Offerings. *Contemporary Accounting Research* 19(1): 49-75.

Copley, P.A., Gaver, J.J., and Gaver, K.M., 1995. Simultaneous Estimation of the Supply and Demand of Differentiated Audits: Evidence from the Municipal Audit Market. *Journal of Accounting Research* 33(1): 137-155.

Craswell, A.T., 1988. The Association between Qualified Opinions and Auditor Switches. *Accounting and Business Research* 19: 23-31.

CSRC, 2005. An Introduction to China's Securities and Futures Markets.

Dahya, J., Karbhari Y, Xiao, J.Z, and Yang M., 2003. The Usefulness of the Supervisory Board Report in China. *Corporate Governance* 11(4): 308-321.

Darrough, M., and Rangan, S., 2005. Do Insiders Manipulate Earnings when They Sell Their Shares in an Initial Public Offering? *Journal of Accounting Research* 43(1): 1-33.

Das, S., and Zhang, H., 2003. Rounding-up in Reported EPS, Behavioral Thresholds, and Earnings Management. *Journal of Accounting and Economics* 35: 31-50.

Datar, S., Feltham, G.A., and Hughes, J.S., 1991. The Role of Audits and Audit Quality in Valuing New Issues. *Journal of Accounting and Economics* 14(1): 3-49.

DeAngelo, L., 1981. Auditor Size and Audit Quality. *Journal of Accounting and Economics* 3(3): 183-199.

DeAngelo, L., 1982. Mandated Successful Efforts and Auditor Choice. *Journal of Accounting and Economics* 4(3): 171-203.

Dechow, P.M., 1994. Accounting Earnings and Cash Flows as Measures of Firm Performance: the Role of Accounting Accruals. *Journal of Accounting and Economics* 18: 3-42.

Dechow, P.M., Richardson, S.A., and Tuna, I., 2003. Why Are Earnings Kinky? An Examination of the Earnings Management Explanation. *Review of Accounting Studies* 8: 355-384.

Dechow, P.M., and Skinner, D.J., 2000. Earnings Management: Reconciling the Views of Accounting Academics, Practitioners, and Regulators. *Accounting Horizons* 14: 235-250.

DeFond, M.L., and Subramanyam, K, 1998. Auditor Changes and Discretionary Accruals. *Journal of Accounting and Economics* 25: 35-67.

DeFond, M.L., Wong, T.J., and Li, S., 2000. The Impact of Improved Auditor Independence on Audit Market Concentration in China. *Journal of Accounting and Economics* 28: 269-305.

Degeorge, F., Patel, J., and Zeckhauser, R., 1999. Earnings Management to Exceed Thresholds. *The Journal of Business* 72: 1-33.

Demski, J.S., Frimor, H., and Sappington, D.E.M., 2004. Efficient Manipulation in a Repeated Setting. *Journal of Accounting Research* 42(1): 31-49.

Dhaliwal, D.S., Gleason, C.A., and Mills, L.F., 2004. Last-Chance Earnings Management: Using the Tax Expense to Meet Analysts' Forecasts. *Contemporary Accounting Research* 21(2): 431-459.

Dopuch, N., Holthausen, R., and Leftwich, R., 1986. Abnormal Stock Returns Associated with Media Disclosures of "Subject to" Qualified Audit Opinions. *Journal of Accounting and Economics* 8(2): 93-117.

Durtschi, C., and Easton, P., 2005. Earnings Management? The Shapes of the Frequency Distributions of Earnings Metrics Are Not Evidence Ipso Facto. *Journal of Accounting Research* 43(4): 557-592.

Dye, R., 1993. Auditing Standards, Legal Liability, and Auditor Wealth. *The Journal of Political Economy* 5: 877-914.

Easton, P., and Pae, J., 2004. Accounting Conservatism and the Relation between Returns and Accounting Data. *Review of Accounting Studies* 9: 495-521

Easton, P., and Zmijewski, M., 1989. Cross-sectional Variation in the Stock Market Response to Accounting Earnings Announcements. *Journal of Accounting and Economics* 11(2/3): 117-141.

Eichenseher, L., and Shields, D., 1986. Corporate Capital Structure and Auditor "Fit". *Working paper*, University of Wisconsin-Madison.

Ertimur, Y., Livnat, J., and Martikainen, M., 2003. Differential Market Reactions to Revenue and Expense Surprises. *Review of Accounting Studies* 8: 185-211.

Fama, E.F., 1980. Agency Problems and the Theory of the Firm. *Journal of Political Economy* 88: 288-307.

Fan, J.P.H., and Wong, T.J., 2002. Corporate Ownership Structure and the Informativeness of Accounting Earnings in East Asia. *Journal of Accounting and Economics* 33: 401-426.

Fan, J.P.H., and Wong, T.J., 2005. Do External Auditors Perform a Corporate Governance Role in Emerging Markets? Evidence from East Asia. *Journal of Accounting Research* 43(1): 35-72.

Felix, W.L., Gramling, A.A., and Maletta, M.J., 2005. The Influence of Nonaudit Service Revenues and Client Pressure on External Auditors' Decisions to Rely on Internal Audit. *Contemporary Accounting Research* 22(1): 31-64.

Felo, A.J., Krishnamurthy, S., and Soleri, S., 2001. Audit Committee Characteristics and the Quality of Financial Reporting: An Empirical Analysis. *Working Paper, Penn State University.*

Feltham, G.A., Hughes, J.S., and Simunic, D.A., 1991. Empirical Assessment of the Impact of Auditor Quality on the Valuation of New Issues. *Journal of Accounting and Economics* 14(4): 375-399.

Ferguson, M.J., Seow, G.S., and Young, D., 2004. Nonaudit Services and Earnings Management: UK Evidence. *Contemporary Accounting Research* 21(4): 813-841.

Fields, T.D., Lys, T.Z., and Vincent, L., 2001. Empirical Research on Accounting Choice. *Journal of Accounting and Economics* 31: 255-307.

Firth, M., 1999. Company Takeovers and the Auditor Choice Decision. *Journal of International Accounting Auditing and Taxation* 8(2): 197-214.

Firth, M., and Smith, A.M.C., 1992. Selection of Auditor Firms by Companies in the New Issue Market. *Applied Economics* 24(2): 247-255.

Firth, M., and Liau-Tan, C.K., 1998. Auditor Quality, Signaling, and the Valuation of Initial Public Offerings. *Journal of Business Finance and Accounting* 25(1/2): 145-165.

Francis, J.R., 2004. What do We Know about Audit Quality. *The British Accounting Review* 36(4): 345-364.

Francis, J.R., and Wilson, E.R., 1988. Auditor Changes: A Joint Test of Theories Relating to Agency Costs and Auditor Differentiation. *The Accounting Review* 63(October): 663-682.

Francis, J.R., and Krishnan, J., 1999. Accounting Accruals and Auditor Reporting Conservatism. *Contemporary Accounting Research* 16(1): 135-166.

Friedlan, J. 1994. Accounting Choices of Issuers of Initial Public Offerings. *Contemporary Accounting Research* 11(1): 1-31.

Geiger, M., Raghunandan, K., and Rama, D.V., 1998. Costs Associated with Going-Concern Modified Audit Opinions: An Analysis of Auditor Changes, Subsequent Opinions, and Client Failures. *Advances in Accounting* 16: 117-139.

Gelb, D.S., and Zarowin, P., 2002. Corporate Disclosure Policy and the Informativeness of Stock Prices. *Review of Accounting Studies* 7: 33-52.

Ghosh, A., Gu, Z., Jain, P.C., 2005. Sustained Earnings and Revenue Growth, Earnings Quality, and Earnings Response Coefficients. *Review of Accounting Studies* 10: 33-57.

Ghosh, A., and Moon, D., 2005. Auditor Tenure and Perceptions of Audit Quality. *The Accounting Review* 80(2): 585-612.

Graham, J.R., Harvey, C.R., Rajgopal, S., 2005. The Economic Implications of Corporate Financial Reporting. Forthcoming, *Journal of Accounting and Economics.*

Han, J.C.Y., and Wang, S., 1998. Political Costs and Earnings Management of Oil Companies during the 1990 Persian Gulf Crisis. *The Accounting Review* 72(1): 103-117.

Haw, I.M., Hu, B., Hwang, L.S., and Wu, W., 2004. Ultimate Ownership, Income Management, and Legal and Extra-Legal Institutions. *Journal of Accounting Research* 42(2): 423-462.

Haw, I.M., Park, K., Qi, D., and Wu, W., 2003. Audit Qualification and Timing of Earnings Announcements: Evidence from China. *Auditing: A Journal of Practice and Theory* 22(2): 121-146.

Haw, I.M., Qi, D., Wu, D., and Wu, W., 2005. Market Consequences of Earnings Management in Response to Security Regulations in China. *Contemporary Accounting Research* 22(1): 95-140.

Hay, D., and Davis, D., 2004. The Voluntary Choice of an Auditor of Any Level of Quality. *Auditing: A Journal of Practice and Theory* 23(2): 37-53.

Hayn, C., 1995. The Information Content of Losses. *Journal of Accounting and Economics* 20: 125-153.

Healy, P., 1985. The Effect of Bonus Schemes on Accounting Choices. *Journal of Accounting and Economics* 7: 85-107.

Healy, P., and Wahlen, J.M., 1999. A Review of the Earnings Management Literature and its Implications for Standard Setting. *Accounting Horizons* 13: 365-383.

Holthausen, R., and Verrecchia, R., 1988. The Effect of Sequential Information Releases on the Variance of Price Changes in an Inter-temporal Multi-asset Market. *Journal of Accounting Research* 26(1): 82-106.

Hribar, P., Jenkins, N.T., and Johnson, W.B., 2006. Stock Repurchases as an Earnings Management Device. *Journal of Accounting and Economics* 41(1/2): 3-27.

Hudaib, M., and Cooke, T.E., 2005. The Impact of Managing Director Changes and Financial Distress on Audit Qualification and Auditor Switching. *Journal of Business Finance & Accounting* 32(9&10): 1703-1739.

Hunton, J.E., Libby, R., and Mazza, C.L., 2006. Financial Reporting Transparency and Earnings Management. *The Accounting Review* 81(1): 135-170.

Iacobucci, E., and Morck, R., 1999. Some of the Causes and Consequences of Corporate Ownership Concentration in Canada (in Concentrated Corporate Ownership). *National Bureau of Economic Research*.

Imhoff, E.A., 2003. Accounting Quality, Auditing, and Corporate Governance. *Accounting Horizons* 17:117-128.

Ireland, J.C., and Lennox, C., 2002. The Large Audit Firm Fee Premium: A Case of Selectivity Bias. *Journal of Accounting, Auditing and Finance* 17(1): 73-98.

Johnson, W. and Lys, T., 1990. The Market for Audit Services: Evidence from Voluntary Auditor Changes. *Journal of Accounting and Economics* 12: 281-308.

Jones, J., 1991. Earnings Management during Import Relief Investigations. *Journal of Accounting Research* 29(2): 193-228.

Joos, P., and Plesko, G.A., 2005. Valuing Loss Firms. *The Accounting Review* 80(3): 847-870.

Kadous, K., 2000. The Effects of Audit Quality and Consequence Severity on Juror Evaluation of Auditor Responsibility for Plaintiff Losses. *The Accounting Review* 75(3): 327-341.

Kanodia, C., and Mukherji, A., 1994. Audit Pricing, Lowballing and Auditor Turnover: A Dynamic Analysis. *The Accounting Review* 69(4): 593-615.

Kasznik, R., and McNichols, M.F., 2002. Does Meeting Earnings Expectations Matter? Evidence from Analyst Forecast Revisions and Share Prices. *Journal of Accounting Research* 40: 737-759.

Kellogg, R., 1984. Accounting Activities, Security Prices, and Class Action Law Suits. *Journal of Accounting and Economics* 6(3): 185-204.

Khurana, I.K., and Raman, K.K., 2004. Litigation Risk and the Financial Reporting Credibility of Big 4 versus Non-Big 4 Audits: Evidence from Anglo-American Countries. *The Accounting Review* 79(2): 473-495.

Kim, J.B., Chung, R., and Firth, M., 2003. Auditor Conservatism, Asymmetric Monitoring, and Earnings Management. *Contemporary Accounting Research* 30(2): 323-359.

Kim, M., and Kross, W., 2005. The Ability of Earnings to Predict Future Operating Cash Flows Has Been Increasing---Not Decreasing. *Journal of Accounting Research* 43(5): 753-780.

King, R.R., and Schwartz, R., 1999. Legal Penalties and Audit Quality: An Experimental Investigation. *Contemporary Accounting Research* 16(4): 685-710.

Klein, A., 2002. Economic Determinants of Audit Committee Independence. *The Accounting Review* 77(2): 435-452.

Klock, M., 1994. The Stock Market Reaction to a Change in Certifying Auditors. *Journal of Accounting, Auditing and Finance* 9: 330-347.

Kothari, S.P., 2001. Capital Markets Research in Accounting. *Journal of Accounting and Economics* 31: 105-231.

Kothari, S.P., and Shanken, J., 2003. Time-Series Coefficient Variation in Value-Relevance Regressions: A Discussion of Core, Guay, and Van Buskirk and New Evidence. *Journal of Accounting and Economics* 34: 69-87.

Krishnan, G.V., 2003. Auditor Quality and the Pricing of Discretionary Accruals. *Auditing: A Journal of Practice & Theory* 22(1): 109-126.

Krishnan, J., 1994. Auditor Switching and Conservatism. *The Accounting Review* 69(1): 200-215.

Krishnan, J., Krishnan, J., and Stephens, R.G., 1996. The Simultaneous Relation Between Auditor Switching and Audit Opinion: An Empirical Analysis. *Accounting and Business Research* 26(3): 224-236.

Krull, L.K., 2004. Permanently Reinvested Foreign Earnings, Taxes, and Earnings Management. *The Accounting Review* 79(3): 745-767.

La Porta, R., Lopez-De-Silanes, F., and Shleifer, A., 1999. Corporate Ownership Around the World. *Journal of Finance* 2: 471-517.

La Porta, R., Lopez-De-Silanes, F., Shleifer, A., and Vishny, R.W., 1998. Law and Finance. *Journal of Political Economy* 106: 1113-1155.

La Porta, R., Lopez-De-Silanes, F., Shleifer, A., and Vishny, R.W., 2002. Investor Protection and Corporate Valuation. *Journal of Finance* 57: 1147-1170.

Lee, H.Y., Mande, V., and Ortman, R., 2004. The Effect of Audit Committee and Board of Director Independence on Auditor Resignation. *Auditing: A Journal of Practice & Theory* 23(2): 131-146.

Lee, P., Stokes, D., Taylor, S., and Walter, T., 2003. The Association between Audit Quality, Accounting Disclosures and Firm-specific Risk: Evidence from Initial Public Offerings. *Journal of Accounting and Public Policy* 22(5): 377-406.

Lennox, C., 1999. Audit Quality and Auditor Size: An Evaluation of Reputation and Deep Pockets Hypotheses. *Journal of Business Finance & Accounting* 26(7/8): 779-805.

Lennox, C., 1999. Non-Audit Fees, Disclosure and Audit Quality. The European Accounting Review 8(2): 239-252.

Lennox, C., 2005. Audit Quality and Executive Officers' Affiliation with CPA Firms. *Journal of Accounting and Economics* 39(2): 201-232.

Lennox, C., 2005. Managing Ownership and Audit Firm Size. *Contemporary Accounting Research* 22(1): 205-227.

Lenz, H., and Ostrowski, M., 2005. Auditor Choice by IPO firms in Germany: Information or Insurance Signaling? *International Journal of Accounting, Auditing and Performance Evaluation* 2(3): 300-320.

Leuz, C., Nanda, D., and Wyosocki, P.D., 2001. Investor Protection and Earnings Management: An International Comparison. *Working Paper.*

Lin, Z.J., and Chen F., 2004. An Empirical Study of Audit 'Expectation Gap' in the People's Republic of China. *International Journal of Auditing* 8(2): 93-115.

Lin, Z.J., Liu, M., and Zhang, X., 2006. The Development of Corporate Governance in China. *Working Paper.*

Lin, Z.J., Tang, Q., and Xiao, J., 2003. An Experimental Study of Users' Responses to Qualified Audit Reports in China. *Journal of International Accounting, Auditing and Taxation* 12: 1-22.

Liu, G.S., and Sun, P., 2005. The Class of Owners and its Impacts on Corporate Performance: a Case of State Shareholding Composition in Chinese Public Corporations. *Corporate Governance* 13(1): 46-71.

Mansi, S.A., Maxwell, W.F., and Miller, D.P., 2004. Does Auditor Quality and Tenure Matter to Investors? Evidence from the Bond Market. *Journal of Accounting Research* 42(4): 755-794.

Marquardt, C.A., and Wiedman, C.I., 2004. How Are Earnings Managed? An Examination of Specific Accruals. *Contemporary Accounting Research* 21(2): 461-491.

Marquardt, C.A., and Wiedman, C.I., 2005. Earnings Management through Transaction Structuring: Contingent Convertible Debt and Diluted Earnings per Share. *Journal of Accounting Research* 43(2): 205-243.

Mayhew, B.W., and Pike, J.E., 2004. Does Investor Selection of Auditors Enhance Auditor Independence? *The Accounting Review* 79(3): 797-822.

Mayhew, B.W., and Schatzberg, J.W., and Sevcik, G.R., 2004. Examining the Role of Auditor Quality and Retained Ownership in IPO Markets: Experimental Evidence. *Contemporary Accounting Research* 21(1): 89-130.

Mayhew, B.W., and Wilkins, M.S., 2003. Audit Firm Industry Specialization as a Differentiation Strategy: Evidence from Fees Charged to Firms Going Public. *Auditing* 22(2): 33-56.

McNichols, M.F., 2003. Discussion of "Why Are Earnings Kinky? An Examination of the Earnings Management Explanation". *Review of Accounting Studies* 8: 385-391.

Monem, R.M., 2003. Earnings Management in Response to the Introduction of the Australian Gold Tax. *Contemporary Accounting Research* 20(4): 747-774.

Menon, K., and Williams, D.D., 2004. Former Audit Partners and Abnormal Accruals. *The Accounting Review* 79(4): 1095-1118.

Myers, J.N., Myers, L.A., and Omer, T.C., 2003. Exploring the Term of the Auditor-Client Relationship and the Quality of Earnings: A Case for Mandatory Auditor Rotation? *The Accounting Review* 78(3): 779-799.

Myers, J.N., Myers, L.A., and Skinner, D.J., 2005. Earnings Momentum and Earnings Management. *Working Paper.*

Nicholes, D. and Smith, D., 1983. Auditor Credibility and Auditor Change. *Journal of Accounting Research*: 534-544.

O'Brian, P.C., 2005. Discussion of Earnings Management through Transaction Structuring: Contingent Convertible Debt and Diluted Earnings per Share. *Journal of Accounting Research* 43(2): 245-250.

Perroti, E.C., and Thadden, E.L., 2005. Dominant Investors and Strategic Transparency. *Journal of Law Economics and Organization* 21(1): 76-102.

Perry, S., and Williams, T., 1994. Earnings Management Preceding Management Buyout Offers. *Journal of Accounting and Economics* 18(2): 157-179.

Philips, J., and Pincus, M., 2003. Earnings Management: New Evidence Based on Deferred Tax Expense. *The Accounting Review* 78(2): 491-521.

Pittman, J.A., and Fortin, S., 2004. Auditor Choice and the Cost of Debt Capital for Newly Public Firms. *Journal of Accounting and Economics* 37(1): 113-136.

Porter, B., Simon, J., and Hatherly, D. 2003. *Principles of External Auditing*. John Wiley & Sons, Chichester.

Reed, B.J., Trombley, M.A., and Dhaliwal, D.S., 2000. Demand for Audit Quality: The Case of Laventhol and Horwath's Auditees. *Journal of Accounting, Auditing & Finance* 15(2): 183-206.

Reynolds, J.K., and Francis, J.R., 2000. Does Size Matter? The Influence of Large Clients on Office-level Auditor Reporting Decisions. *Journal of Accounting and Economics* 30: 375-400.

Rogers, J.L., and Stocken, P.C., 2005. Credibility of Management Forecasts. *The Accounting Review* 80(4): 1233-1260.

Rosner, R.L., 2003. Earnings Manipulation in Failing Firms. *Contemporary Accounting Research* 20(2): 361-408.

Ryan, S.G., and Zarowin, P.A., 2003. Why has the Contemporaneous Linear Returns-Earnings Relation Declined? *The Accounting Review* 78(2): 523-553.

Schatzberg, J.W., and Sevcik, G.R., 1994. A Multiperiod Model and Experimental Evidence of Independence and "Lowballing". *Contemporary Accounting Research* 11(1): 137-174.

Schauer, P.C., 2002. The Effects of Industry Specialization on Audit Quality: An Examination Using Bid-Ask Spreads. *Journal of Accounting and Finance Research* 10(1): 76-86.

Sheng, J., 2004. The Legal Supervisory Mechanism of Chinese Listed Companies. *Working Paper*.

Shleifer, A., and Vishny, R.W., 1997. A Survey of Corporate Governance. *The Journal of Finance* 52(2): 737-783.

Schrand, C.M., and Wong, M.H.F., 2003. Earnings Management Using the Valuation Allowance for Deferred Tax Assets under SFAS No. 109. *Contemporary Accounting Research* 20(3): 579-611.

Schwartz, K.B., and Menon, K., 1985. Auditor Switch by Failing Firms. *The Accounting Review* 60(2): 248-261.

Simunic, D.A., Stein, M.T., 1987. Product Differentiation in Auditing: Auditor Choice in the Market for Unseasoned New Issues. *Canadian Certified General Accountants' Research Foundation*, Vancouver, BC.

Stice, J., 1991. Using Financial and Market Information to Identify Pre-engagement Factors Associated with Lawsuits against Auditors. *The Accounting Review* 66(3): 516-533.

Teoh, S.H., 1992. Auditor Independence, Dismissal Threats, and the Market Reaction to Auditor Switches. *Journal of Accounting Research* 30(1): 1-26.

Teoh, S.H., and Wong, T.J., 1993. Perceived Auditor Quality and the Earnings Response Coefficient. *The Accounting Review* 68: 346-366.

Thornton, D.B., and Moore, G., 1993. Auditor Choice and Audit Fee Determinants. *Journal of Business Finance and Accounting* 20(3): 333-348.

Titman, S., and Trueman, B., 1986. Information Quality and the Valuation of New Issues. *Journal of Accounting and Economics* 8(2): 159-172.

Wallace, W., 1987. The Economic Role of the Audit in Free and Regulated Markets: A Review. *Research in Accounting Regulation* 1:1-34.

Watkins, A.L., Hillison, W., and Morecroft, S.E., 2004. Audit Quality: A Synthesis of Theory and Empirical Evidence. *Journal of Accounting Literature* 23: 153-193.

Watts, R.L., Zimmerman, J.L., 1981. The Markets for Independence and Independent Auditors. *Working Paper*, University of Rochster.

Watts, R.L., Zimmerman, J.L., 1986. *Positive Accounting Theory*. Prentice-Hall, Englewood Cliffs, NJ.

Watts, R.L., Zimmerman, J.L., 1990. Positive Accounting Theory: A Ten Year Perspective. *The Accounting Review* 65(1): 131-156.

Willenborg, M., 1999. Empirical Analysis of the Economic Demand for Auditing in the Initial Public Offerings Market. *Journal of Accounting Research* 37(1): 225-239.

Woodward, D.G., 1997. Book Reviews: FEE 1994 Investigation of Emerging Accounting Areas. *The British Accounting Review* 29(1): 90-93.

Xiang, B., 1998. Institutional Factors Influencing China's Accounting Reforms and Standards. *Accounting Horizons* 12(2): 105-119.

Xiao, J.Z., Zhang, Y., and Xie, Z., 2000. The Making of Independent Auditing Standards in China. *Accounting Horizons* 14(1): 69-89.

Xiao, J.Z., Dahya, J., and Lin, Z., 2004. A Grounded Theory Exposition of the Role of the Supervisory Board in China. *British Journal of Management* 15: 39-63.

Yang, L., Dunk, A., Kilgore, A., Tang, Q., and Lin, Z.J., 2003. Auditor Independence Issues in China. *Managerial Finance* 29(12): 57-64.

Yang, Y., 1995. On the Public Image of the CPA. *CPA Newsletter* 9: 11-13.

Table 1: Revenues and rankings of Top 10 auditors in China in 2002-2004

Auditors	2002		2003		2004	
	Revenues	Rank	Revenues	Rank	Revenues	Rank
PwC Zhongtian	766	1	902	1	1247	1
KPMG Huazhen	334	2	432	2	716	2
Deloitte Huayong	292	3	376	3	658	3
EY Huaming	246	4	329	4	628	4
Lixin Changjiang	101	5	114	5	153	5
Yuehua	83	6	86	6	101	8
Xinyongzhonghe	65	9	85	7	121	6
Beijing Jingdu	65	8	66	11	75	10
Jiangsu Gongzheng	48	12	68	10	73	12
EY Dahua	55	11	59	12	62	17

Notes:

1) Annual audit revenues are in RMB million.

2) Top 10 auditors are the ten largest auditors based on the average audit revenues of Year 2002-04. Therefore, Top 10 auditors may not be the same as the ten largest auditors in a specific year of 2002, 2003 or 2004.

Table 2: Description of data for tests of H1 to H3

Panel A1: Sample selection

IPO firms during 2001-2004	317
Less: Firms with missing data	121
Financial, transportation, and utility firms	12
Final sample	184

Panel A2: Sample distribution by sector and year

Sector	2001	2002	2003	2004	Total
Industry	34	30	28	27	119
Commerce	5	6	5	5	21
Property	2	3	3	1	9
Conglomerate	10	9	8	8	35
Total	51	48	44	41	184

Panel B: Descriptive statistics of variables

Variable	N	Mean	Median	MIN	MAX	STD
AUD	184	0.23	0	0	1	0.42
LSH (%)	184	46.09	48.83	1.13	71.02	16.99
SB (#)	184	4.41	3	1	13	2.08
CEOCHR	184	0.11	0	0	1	0.31
LNASSET	184	20.12	19.95	18.23	24.47	1.05
ATR	184	0.90	0.77	0.09	3.77	0.58
ROA	184	0.10	0.09	0.02	0.25	0.05
CURR	184	1.40	1.27	0.36	5.30	0.77
DA	184	0.54	0.56	0.14	0.91	0.13
ABBETA	184	1.09	1.03	0.00	4.81	0.63

Panel C: Correlation coefficient matrix

	AUD	LSH	SB	CEOCHR	LNASSET	ATR	ROA	CURR	DA	ABBETA
AUD	1.000***									
LSH	-.301***	1.000***								
SB	.199***	-.030	1.000***							
CEOCHR	-.107*	.000	.150**	1.000***						
LNASSET	.187***	.095*	.340***	-.093	1.000***					
ATR	.127**	.119*	-.122**	.058	-.146**	1.000***				
ROA	.055	-.037	-.125**	.100*	-.367***	.295***	1.000***			
CURR	.094	-.001	.020	.077	-.032	-.164**	-.032	1.000***		
DA	.119**	.059	-.051	-.090	.026	.289***	-.090	-.666***	1.000***	
ABBETA	-.016	.021	-.064	.157**	-.033	-.021	-.072	.022	.045	1.000***

***, **, and * denote significance at the 1%, 5%, and 10% levels, respectively.

The variables are defined as below:

AUD	=1 if the auditor is a Top 10 auditor; 0 otherwise
LSH	=the largest owner's shareholding as a percentage of total shares
SB	=number of SB members
CEOCHR	=1 if the CEO also holds the position of the BoD chairman; 0 otherwise
LNASSET	=log of total assets
ATR	=asset turnover ratio, calculated as sales divided by total assets
ROA	=earnings after interest and taxation divided by total assets
CURR	=current assets divided by total assets
DA	=total liabilities divided by total assets
ABBETA	=the absolute value of beta to proxy for firm risk

Table 3: Internal corporate governance mechanism and choosing a Top 10 auditor
--- Test of H1 to H3

$AUD = \beta_0 + \beta_1 LSH + \beta_2 SB + \beta_3 CEOCHR + \beta_4 LNASSET + \beta_5 ATR + \beta_6 ROA + \beta_7 CURR + \beta_8 DA + \beta_9 ABBETA + \varepsilon$ (1)

	Intercept	LSH	SB	CEOCHR	LNASSET	ATR	ROA	CURR	DA	ABBETA
	β_0	β_1	β_2	β_3	β_4	β_5	β_6	β_7	β_8	β_9
Predictions	?	-	+	-	+	+	+	?	?	?
Coeff.	-15.344	-.057	.268	-1.589	.623	.751	8.500	-.055	2.288	.110
Wald	9.291***	19.018***	5.671**	3.311*	7.651***	4.573**	3.197*	.019	1.100	.087

Notes:

1) The Wald-Wolfowitz test examines whether the mean (median) of each variable for Top 10 clients equals to that of non Top-10 clients. ***, **, and * denote significance at the 1%, 5%, and 10% levels, respectively.

2) N = 184, Chi-square = 45.233, and Pseudo R-square = .331.

The variables are defined as below:

AUD	=1 if the auditor is a Top 10 auditor; 0 otherwise
LSH	=the largest owner's shareholding as a percentage of total shares
SB	=number of SB members
CEOCHR	=1 if the CEO also holds the position of the BoD chairman; 0 otherwise
LNASSET	=log of total assets
ATR	=asset turnover ratio, calculated as sales divided by total assets
ROA	=earnings after interest and taxation divided by total assets
CURR	=current assets divided by total assets
DA	=total liabilities divided by total assets
ABBETA	=the absolute value of beta to proxy for firm risk

Table 3-1: Internal corporate governance mechanism and choosing a Top 10 auditor
--- Test of H1 to H3 (sensitivity: ten percent cutoff)

$$AUD = \beta_0 + \beta_1 LSH + \beta_2 SB + \beta_3 CEOCHR + \beta_4 LNASSET + \beta_5 ATR + \beta_6 ROA + \beta_7 CURR + \beta_8 DA + \beta_9 ABBETA + \varepsilon \qquad (1)$$

	Intercept	LSH	SB	CEOCHR	LNASSET	ATR	ROA	CURR	DA	ABBETA
	β_0	β_1	β_2	β_3	β_4	β_5	β_6	β_7	β_8	β_9
Predictions	?	-	+	-	+	+	+	?	?	?
Coeff.	-15.887	-.044	.272	-1.493	.591	.741	8.405	.051	2.866	.176
Wald[a]	9.214***	9.592***	6.032**	2.908*	6.422**	4.493**	3.059*	.016	1.564	.221

Notes:

1) To ensure effective control, this sensitivity test excludes observations in which the largest shareholders hold less than 10% of total shares.

2) The Wald-Wolfowitz test examines whether the mean (median) of each variable for Top 10 clients equals to that of non Top-10 clients. ***, **, and * denote significance at the 1%, 5%, and 10% levels, respectively.

3) N = 179, Chi-square = 36.233, and Pseudo R-square = .287.

The variables are defined as below:

AUD	=1 if the auditor is a Top 10 auditor; 0 otherwise
LSH	=the largest owner's shareholding as a percentage of total shares
SB	=number of SB members
CEOCHR	=1 if the CEO also holds the position of the BoD chairman; 0 otherwise
LNASSET	=log of total assets
ATR	=asset turnover ratio, calculated as sales divided by total assets
ROA	=earnings after interest and taxation divided by total assets
CURR	=current assets divided by total assets
DA	=total liabilities divided by total assets
ABBETA	=the absolute value of beta to proxy for firm risk

Table 4: Description of data for H4 to H6

Panel A1: Sample selection

Firms switching auditors during 2001-04	316
Less: Firms with missing data	47
Financial, transportation, and utility firms	11
Firms switching auditors for more than once during 2001-04	25
Firms switching auditors among Top 10 auditors or among non-Top 10 auditors	171
Final sample	62

Panel A2: Sample distribution by sector and year

Sector	2001	2002	2003	2004	Total
Industry	11	10	8	4	33
Commerce	2	2	3	2	9
Property	2	3	2	2	9
Conglomerate	2	3	3	3	11
Total	17	18	16	11	62

Panel B: Descriptive statistics of variables

Variable	N	Mean	Median	MIN	MAX	STD
SD	62	0.42	0	0	1	0.50
LSH (%)	62	47.82	47.39	6.09	83.33	18.30
SB (#)	62	4.55	3.5	1	12	2.16
CEOCHR	62	0.11	0	0	1	0.32
LNASSET	62	21.14	21.11	19.47	24.60	0.98
LEV	62	0.10	0.07	0.00	0.43	0.11
MB	62	4.27	3.74	1.27	11.53	2.16
LOSS	62	0.10	0	0	1	0.30
OPI	62	0.16	0	0	1	0.37

Panel C: Correlation coefficient matrix

Variable	SD	LSH	SB	CEOCHR	LNASSET	LEV	MB	LOSS	OPI
SD	1.000***								
LSH	.141**	1.000***							
SB	.194*	-.096	1.000***						
CEOCHR	.317***	-.279**	.194*	1.000***					
LNASSET	-.226**	.417***	-.120	-.232**	1.000***				
LEV	.064	-.093	-.059	-.213**	.230**	1.000***			
MB	.096	-.260**	.208*	-.208*	-.390***	-.064	1.000***		
LOSS	-.168	-.092	-.135	-.117	-.137	-.059	.128	1.000***	
OPI	-.095	-.073	.092	-.018	-.067	-.078	.339***	.598***	1.000***

***, **, and * denote significance at the 1%, 5%, and 10% levels, respectively.

The variables are defined as below:

SD	=1 if the firm switches from a Top 10 auditor to a non-Top 10 auditor; 0 if the firm switches from a non-Top 10 auditor to a Top-10 auditor
LSH	=the largest owner's shareholding as a percentage of total shares
SB	=number of SB members
CEOCHR	=1 if the CEO also holds the position of the BoD chairman; 0 otherwise
LNASSET	=log of total assets at the end of the previous year
LEV	=long-term liabilities divided by total assets at the end of the previous year
MB	=market-to-book ratio at the end of the previous year, calculated as the market value of stocks divided by the book value
LOSS	=1 if the firm experiences a net loss for the previous year; 0 otherwise
OPI	=1 if the firm receives an unclean auditor opinion for the previous year; 0 otherwise

Table 5: Internal corporate governance mechanism and switching to a smaller auditor --- Test of H4 to H6

$$SD = \gamma_0 + \gamma_1 LSH + \gamma_2 SB + \gamma_3 CEOCHR + \gamma_4 LNASSET + \gamma_5 LEV + \gamma_6 MB + \gamma_7 LOSS + \gamma_8 OPI + \varepsilon \quad (2)$$

	Intercept	LSH	SB	CEOCHR	LNASSET	LEV	MB	LOSS	OPI
	γ_0	γ_1	γ_2	γ_3	γ_4	γ_5	γ_6	γ_7	γ_8
Predictions	?	+	-	+	-	+	+	+	+
Coeff.	14.058	.046	.207	2.955	-.871	4.966	.048	.023	-1.629
Wald[a]	2.449	4.467**	1.498	5.133**	3.898**	3.039**	.066	.000	1.497

Notes:

1) In this sensitivity test we use a more strict measure to proxy for SD and SU. Here SD is defined as switching from a Top 10 auditor to a non-Top 10 auditor and SU as switching from a non-Top 10 auditor to a Top 10 auditor.

2) The Wald-Wolfowitz test examines whether the mean (median) of each variable for Top 10 clients equals to that of non Top-10 clients. ***, **, and * denote significance at the 1%, 5%, and 10% levels, respectively.

3) N = 62, Chi-square = 19.576, and Pseudo R-square = .364.

The variables are defined as below:

SD	=1 if the firm switches from a Top 10 auditor to a non-Top 10 auditor; 0 if the firm switches from a non-Top 10 auditor to a Top-10 auditor
LSH	=the largest owner's shareholding as a percentage of total shares
SB	=number of SB members
CEOCHR	=1 if the CEO also holds the position of the BoD chairman; 0 otherwise
LNASSET	=log of total assets at the end of the previous year
LEV	=long-term liabilities divided by total assets at the end of the previous year
MB	=market-to-book ratio at the end of the previous year, calculated as the market value of stocks divided by the book value
LOSS	=1 if the firm experiences a net loss for the previous year; 0 otherwise
OPI	=1 if the firm receives an unclean auditor opinion for the previous year; 0 otherwise

Table 5-1: Internal corporate governance mechanism and switching to a smaller auditor --- Test of H4 to H6 (sensitivity: including switches among Top 10 or among non-Top 10)

$$SD = \gamma_0 + \gamma_1 LSH + \gamma_2 SB + \gamma_3 CEOCHR + \gamma_4 LNASSET + \gamma_5 LEV + \gamma_6 MB + \gamma_7 LOSS + \gamma_8 OPI + \varepsilon$$
(2)

	Intercept	LSH	SB	CEOCHR	LNASSET	LEV	MB	LOSS	OPI
	γ_0	γ_1	γ_2	γ_3	γ_4	γ_5	γ_6	γ_7	γ_8
Predictions	?	+	-	+	-	+	+	+	+
Coeff.	5.922	.021	.091	1.086	-.411	3.633	.127	.304	-.159
Wald[a]	2.169	6.229**	1.175	3.287*	4.673**	4.332**	4.158**	.399	.131

Notes:

1) The Wald-Wolfowitz test examines whether the mean (median) of each variable for Top 10 clients equals to that of non Top-10 clients. ***, **, and * denote significance at the 1%, 5%, and 10% levels, respectively.

2) N = 233, Chi-square = 28.308, and Pseudo R-square = .154.

The variables are defined as below:

SD	=1 if the firm switches from a Top 10 auditor to a non-Top 10 auditor; 0 if the firm switches from a non-Top 10 auditor to a Top-10 auditor
LSH	=the largest owner's shareholding as a percentage of total shares
SB	=number of SB members
CEOCHR	=1 if the CEO also holds the position of the BoD chairman; 0 otherwise
LNASSET	=log of total assets at the end of the previous year
LEV	=long-term liabilities divided by total assets at the end of the previous year
MB	=market-to-book ratio at the end of the previous year, calculated as the market value of stocks divided by the book value
LOSS	=1 if the firm experiences a net loss for the previous year; 0 otherwise
OPI	=1 if the firm receives an unclean auditor opinion for the previous year; 0 otherwise

Table 6: Description of data for H7a, H8, and H9

Panel A1: Sample selection

Firm-years with positive abnormal earnings for 2002-2004	1961
Less: Firm-years with missing data	377
Financial, transportation, and utility firm-years	62
Firms changing CEOs during 2002-2004	225
Firms switching auditors for more than once during 2001-04	13
Final sample	1284

Panel A2: Sample distribution by sector and year

Sector	2002	2003	2004	Total
Industry	189	201	429	819
Commerce	29	42	70	141
Property	12	26	46	84
Conglomerate	55	71	114	240
Total	285	340	659	1284

Panel B: Descriptive statistics of variables

Variable	N	Mean	Median	MIN	MAX	STD
CAR	1284	-.006	-.005	-.200	.198	.043
ΔE	1284	.025	.008	.000	.708	.056
T10	1284	.234	0	0	1	.424
SU	1284	.112	0	0	1	.316
SD	1284	.070	0	0	1	.255
OPI	1284	.040	0	0	1	.197
MB	1284	3.692	2.409	.797	252.82	12.351
ABBETA	1284	1.033	1.027	.001	3.949	.423

Panel C: Correlation coefficient matrix

Variable	CAR	ΔE	T10	SU	SD	OPI	MB	ABBETA
CAR	1.000***							
ΔE	-.026	1.000***						
T10	.032	-.041*	1.000***					
SU	.170***	.039*	.106***	1.000***				
SD	-.029	.044*	-.094***	-.088***	1.000***			
OPI	-.025	.273***	.008	.040*	.114***	1.000***		
MB	-.003	.057**	-.040*	-.006	.050**	.218***	1.000***	
ABBETA	-.063**	.130***	.018	-.049**	.020	.077***	-.017	1.000***

***, **, and * denote significance at the 1%, 5%, and 10% levels, respectively.

The variables are defined as below:

CAR	=market-adjusted abnormal returns accumulated over the two trading days (0, +1), where 0 is the annual earnings announcement date, adjusted for dividends and stock rights
ΔE	=unexpected annual earnings, calculated as earnings of this year minus earnings of last year and then divided by market value of equity
T10	=1 if the firm is audited by a Top 10 auditor, and 0 otherwise
SU	=1 if the firm switches to a larger auditor in the current or the last year, and 0 otherwise
SD	=1 if the firm switches to a smaller auditor in the current or the last year, and 0 otherwise
OPI	=1 if the firm receives an unclean auditor's opinion in the current year, and 0 otherwise
MB	=market-to-book ratio at the end of the previous year, calculated as the market value of the firm's stock divided by its book value
ABBETA	=the absolute value of beta to proxy for firm risk

Table 7: Description of data for H7b, H8, and H9

Panel A1: Sample selection

Firm-years with negative abnormal earnings for 2002-2004	1626
Less: Firm-years with missing data	344
Financial, transportation, and utility firm-years	37
Firms changing CEOs during 2002-2004	252
Firms switching auditors for more than once during 2001-04	12
Final sample	981

Panel A2: Sample distribution by sector and year

Sector	2002	2003	2004	Total
Industry	197	123	306	626
Commerce	38	23	58	119
Property	15	13	22	50
Conglomerate	62	35	89	186
Total	312	194	475	981

Panel B: Descriptive statistics of variables

Variable	N	Mean	Median	MIN	MAX	STD
CAR	981	-.010	-.006	-.201	.201	.045
ΔE	981	-.038	-.010	-.854	-.000	.086
T10	981	.211	0	0	1	.408
SU	981	.112	0	0	1	.316
SD	981	.085	0	0	1	.278
OPI	981	.107	0	0	1	.309
MB	981	3.845	2.256	.669	482.45	18.496
ABBETA	981	1.055	1.050	.002	3.095	.413

Panel C: Correlation coefficient matrix

Variable	CAR	ΔE	T10	SU	SD	OPI	MB	ABBETA
CAR	1.000***							
ΔE	.084***	1.000***						
T10	.050*	-.001	1.000***					
SU	.081***	.010	.188***	1.000***				
SD	-.060**	-.075***	-.112***	-.108***	1.000***			
OPI	-.063**	-.465***	-.001	.023	.061**	1.000***		
MB	-.036	-.107***	-.030	-.014	.061**	.032	1.000***	
ABBETA	-.071**	-.145***	.011	.023	-.011	.071**	.079***	1.000***

***, **, and * denote significance at the 1%, 5%, and 10% levels, respectively.

The variables are defined as below:

CAR	=market-adjusted abnormal returns accumulated over the two trading days (0, +1), where 0 is the annual earnings announcement date, adjusted for dividends and stock rights
ΔE	=unexpected annual earnings, calculated as earnings of this year minus earnings of last year and then divided by market value of equity
T10	=1 if the firm is audited by a Top 10 auditor, and 0 otherwise
SU	=1 if the firm switches to a larger auditor in the current or the last year, and 0 otherwise
SD	=1 if the firm switches to a smaller auditor in the current or the last year, and 0 otherwise
OPI	=1 if the firm receives an unclean auditor's opinion in the current year, and 0 otherwise
MB	=market-to-book ratio at the end of the previous year, calculated as the market value of the firm's stock divided by its book value
ABBETA	=the absolute value of beta to proxy for firm risk

Table 8: Market implications of auditor choice and auditor switch for positive ΔE
--- Test of H7a, H8 and H9

$$CAR = \delta_0 + \delta_1\Delta E + \delta_2 T10*\Delta E + \delta_3 SU*\Delta E + \delta_4 SD*\Delta E + \delta_5 OPI + \delta_6 OPI*\Delta E + \delta_7 MB*\Delta E + \delta_8 ABBETA*\Delta E + \varepsilon \qquad (3)$$

	Intercept δ_0	ΔE δ_1	T10*ΔE δ_2	SU*ΔE δ_3	SD*ΔE δ_4	OPI δ_5	OPI*ΔE δ_6	MB*ΔE δ_7	ABBETA*ΔE δ_8
Predictions	?	+	+	+	-	-	-	+	-
Coeff.	-.005	.083	.198	.266	-.215	-.001	-.006	.002	-.138
t	-3.994***	1.657*	3.817***	4.588***	-3.108***	-.181	-.091	1.729*	-3.551***

Notes:

1) ***, **, and * denote significance at the 1%, 5%, and 10% levels, respectively.

2) N= 1284, F-statistic = 7.287, and Adjusted R-square = .038.

The variables are defined as below:

CAR	=market-adjusted abnormal returns accumulated over the two trading days (0, +1), where 0 is the annual earnings announcement date, adjusted for dividends and stock rights
ΔE	=unexpected annual earnings, calculated as earnings of this year minus earnings of last year and then divided by market value of equity
T10	=1 if the firm is audited by a Top 10 auditor, and 0 otherwise
SU	=1 if the firm switches to a larger auditor in the current or the last year, and 0 otherwise
SD	=1 if the firm switches to a smaller auditor in the current or the last year, and 0 otherwise
OPI	=1 if the firm receives an unclean auditor's opinion in the current year, and 0 otherwise
MB	=market-to-book ratio at the end of the previous year, calculated as the market value of the firm's stock divided by its book value
ABBETA	=the absolute value of beta to proxy for firm risk

Table 9: Market implications of auditor choice and auditor switch for negative ΔE
--- Test of H7b, H8 and H9

$$CAR = \delta_0 + \delta_1\Delta E + \delta_2 T10*\Delta E + \delta_3 SU*\Delta E + \delta_4 SD*\Delta E + \delta_5 OPI + \delta_6 OPI*\Delta E + \delta_7 MB*\Delta E + \delta_8 ABBETA*\Delta E + \varepsilon \qquad (3)$$

	Intercept	ΔE	T10*ΔE	SU*ΔE	SD*ΔE	OPI	OPI*ΔE	MB*ΔE	ABBETA*ΔE
	δ_0	δ_1	δ_2	δ_3	δ_4	δ_5	δ_6	δ_7	δ_8
Predictions	?	+	-	-	+	-	+	?	?
Coeff.	-.009	.004	-.148	-.222	.106	-.000	.083	.000	.019
t	-5.309***	.069	-3.885***	-4.064***	2.308**	-.013	1.958*	.406	.487

Notes:

1) ***, **, and * denote significance at the 1%, 5%, and 10% levels, respectively.

2) N = 981, F-statistic = 7.712, and Adjusted R-square = .052.

The variables are defined as below:

CAR	=market-adjusted abnormal returns accumulated over the two trading days (0, +1), where 0 is the annual earnings announcement date, adjusted for dividends and stock rights
ΔE	=unexpected annual earnings, calculated as earnings of this year minus earnings of last year and then divided by market value of equity
T10	=1 if the firm is audited by a Top 10 auditor, and 0 otherwise
SU	=1 if the firm switches to a larger auditor in the current or the last year, and 0 otherwise
SD	=1 if the firm switches to a smaller auditor in the current or the last year, and 0 otherwise
OPI	=1 if the firm receives an unclean auditor's opinion in the current year, and 0 otherwise
MB	=market-to-book ratio at the end of the previous year, calculated as the market value of the firm's stock divided by its book value
ABBETA	=the absolute value of beta to proxy for firm risk

Table 10-1: Market implications of auditor choice and auditor switch for positive ΔE in 2002 --- Test of H7a, H8 and H9

$$CAR = \delta_0 + \delta_1\Delta E + \delta_2 T10*\Delta E + \delta_3 SU*\Delta E + \delta_4 SD*\Delta E + \delta_5 OPI + \delta_6 OPI*\Delta E + \delta_7 MB*\Delta E + \delta_8 ABBETA*\Delta E + \varepsilon \qquad (3)$$

	Intercept δ_0	ΔE δ_1	T10*ΔE δ_2	SU*ΔE δ_3	SD*ΔE δ_4	OPI δ_5	OPI*ΔE δ_6	MB*ΔE δ_7	ABBETA*ΔE δ_8
Predictions	?	+	+	+	-	-	-	+	-
Coeff.	-.001	.213	.283	.226	-.233	.003	.105	-.002	-.173
t	-.327	2.593***	2.041**	2.119**	-1.978**	.284	.716	-1.436	-2.879***

Notes:

1) In this sensitivity test we run the regression using yearly data of 2002 with positive abnormal earnings.

2) ***, **, and * denote significance at the 1%, 5%, and 10% levels, respectively.

3) N = 285, F-statistic = 4.187, and Adjusted R-square = .082.

The variables are defined as below:

CAR	=market-adjusted abnormal returns accumulated over the two trading days (0, +1), where 0 is the annual earnings announcement date, adjusted for dividends and stock rights
ΔE	=unexpected annual earnings, calculated as earnings of this year minus earnings of last year and then divided by market value of equity
T10	=1 if the firm is audited by a Top 10 auditor, and 0 otherwise
SU	=1 if the firm switches to a larger auditor in the current or the last year, and 0 otherwise
SD	=1 if the firm switches to a smaller auditor in the current or the last year, and 0 otherwise
OPI	=1 if the firm receives an unclean auditor's opinion in the current year, and 0 otherwise
MB	=market-to-book ratio at the end of the previous year, calculated as the market value of the firm's stock divided by its book value
ABBETA	=the absolute value of beta to proxy for firm risk

Table 10-2: Market implications of auditor choice and auditor switch for positive ΔE in 2003 --- Test of H7a, H8 and H9

$$CAR = \delta_0 + \delta_1\Delta E + \delta_2 T10 * \Delta E + \delta_3 SU * \Delta E + \delta_4 SD * \Delta E + \delta_5 OPI + \delta_6 OPI * \Delta E + \delta_7 MB * \Delta E + \delta_8 ABBETA * \Delta E + \varepsilon \qquad (3)$$

	Intercept δ_0	ΔE δ_1	T10*ΔE δ_2	SU*ΔE δ_3	SD*ΔE δ_4	OPI δ_5	OPI*ΔE δ_6	MB*ΔE δ_7	ABBETA*ΔE δ_8
Predictions	?	+	+	+	-	-	-	+	-
Coeff.	.001	.066	.366	.317	-.372	-.001	-.015	-.001	-.142
t	.272	.337	2.186**	2.042**	-2.025**	-.046	-.069	-.122	-.834

Notes:

1) In this sensitivity test we run the regression using yearly data of 2003 with positive abnormal earnings.

2) ***, **, and * denote significance at the 1%, 5%, and 10% levels, respectively.

3) N = 340, F-statistic = 3.481, and Adjusted R-square = .055.

The variables are defined as below:

CAR	=market-adjusted abnormal returns accumulated over the two trading days (0, +1), where 0 is the annual earnings announcement date, adjusted for dividends and stock rights
ΔE	=unexpected annual earnings, calculated as earnings of this year minus earnings of last year and then divided by market value of equity
T10	=1 if the firm is audited by a Top 10 auditor, and 0 otherwise
SU	=1 if the firm switches to a larger auditor in the current or the last year, and 0 otherwise
SD	=1 if the firm switches to a smaller auditor in the current or the last year, and 0 otherwise
OPI	=1 if the firm receives an unclean auditor's opinion in the current year, and 0 otherwise
MB	=market-to-book ratio at the end of the previous year, calculated as the market value of the firm's stock divided by its book value
ABBETA	=the absolute value of beta to proxy for firm risk

Table 10-3: Market implications of auditor choice and auditor switch for positive ΔE in 2004 --- Test of H7a, H8 and H9

$$CAR = \delta_0 + \delta_1\Delta E + \delta_2 T10^*\Delta E + \delta_3 SU^*\Delta E + \delta_4 SD^*\Delta E + \delta_5 OPI + \delta_6 OPI^*\Delta E + \delta_7 MB^*\Delta E + \delta_8 ABBETA^*\Delta E + \varepsilon \qquad (3)$$

	Intercept	ΔE	T10*ΔE	SU*ΔE	SD*ΔE	OPI	OPI*ΔE	MB*ΔE	ABBETA*ΔE
	δ_0	δ_1	δ_2	δ_3	δ_4	δ_5	δ_6	δ_7	δ_8
Predictions	?	+	+	+	-	-	-	+	-
Coeff.	-.011	.193	.144	.185	-.201	-.002	-.014	.003	-.216
t	-5.158***	2.130**	2.093**	2.066**	-1.983**	-.149	-.148	1.544	-3.063***

Notes:

1) In this sensitivity test we run the regression using yearly data of 2004 with positive abnormal earnings.

2) ***, **, and * denote significance at the 1%, 5%, and 10% levels, respectively.

3) N = 659, F-statistic = 3.180, and Adjusted R-square = .026.

The variables are defined as below:

CAR	=market-adjusted abnormal returns accumulated over the two trading days (0, +1), where 0 is the annual earnings announcement date, adjusted for dividends and stock rights
ΔE	=unexpected annual earnings, calculated as earnings of this year minus earnings of last year and then divided by market value of equity
T10	=1 if the firm is audited by a Top 10 auditor, and 0 otherwise
SU	=1 if the firm switches to a larger auditor in the current or the last year, and 0 otherwise
SD	=1 if the firm switches to a smaller auditor in the current or the last year, and 0 otherwise
OPI	=1 if the firm receives an unclean auditor's opinion in the current year, and 0 otherwise
MB	=market-to-book ratio at the end of the previous year, calculated as the market value of the firm's stock divided by its book value
ABBETA	=the absolute value of beta to proxy for firm risk

Table 11-1: Market implications of auditor choice and auditor switch for negative ΔE in 2002 --- Test of H7b, H8 and H9

$$CAR = \delta_0 + \delta_1\Delta E + \delta_2 T10^*\Delta E + \delta_3 SU^*\Delta E + \delta_4 SD^*\Delta E + \delta_5 OPI + \delta_6 OPI^*\Delta E + \delta_7 MB^*\Delta E + \delta_8 ABBETA^*\Delta E + \varepsilon \qquad (3)$$

	Intercept δ_0	ΔE δ_1	T10*ΔE δ_2	SU*ΔE δ_3	SD*ΔE δ_4	OPI δ_5	OPI*ΔE δ_6	MB*ΔE δ_7	ABBETA*ΔE δ_8
Predictions	?	+	-	-	+	-	+	?	?
Coeff.	-.003	.012	-.433	-.381	.402	-.002	.015	-.000	.076
t	-1.440	.076	-2.045**	-2.851***	2.145**	-.317	.119	-.025	.629

Notes:

1) In this sensitivity test we run the regression using yearly data of 2002 with negative abnormal earnings.

2) ***, **, and * denote significance at the 1%, 5%, and 10% levels, respectively.

3) N = 312, F-statistic = 3.257, and Adjusted R-square = .055.

The variables are defined as below:

CAR	=market-adjusted abnormal returns accumulated over the two trading days (0, +1), where 0 is the annual earnings announcement date, adjusted for dividends and stock rights
ΔE	=unexpected annual earnings, calculated as earnings of this year minus earnings of last year and then divided by market value of equity
T10	=1 if the firm is audited by a Top 10 auditor, and 0 otherwise
SU	=1 if the firm switches to a larger auditor in the current or the last year, and 0 otherwise
SD	=1 if the firm switches to a smaller auditor in the current or last year, and 0 otherwise
OPI	=1 if the firm receives an unclean auditor's opinion in the current year, and 0 otherwise
MB	=market-to-book ratio at the end of the previous year, calculated as the market value of the firm's stock divided by its book value
ABBETA	=the absolute value of beta to proxy for firm risk

Table 11-2: Market implications of auditor choice and auditor switch for negative ΔE in 2003 --- Test of H7b, H8 and H9

$$CAR = \delta_0 + \delta_1\Delta E + \delta_2 T10*\Delta E + \delta_3 SU*\Delta E + \delta_4 SD*\Delta E + \delta_5 OPI + \delta_6 OPI*\Delta E + \delta_7 MB*\Delta E + \delta_8 ABBETA*\Delta E + \varepsilon \qquad (3)$$

	Intercept	ΔE	T10*ΔE	SU*ΔE	SD*ΔE	OPI	OPI*ΔE	MB*ΔE	ABBETA*ΔE
	δ_0	δ_1	δ_2	δ_3	δ_4	δ_5	δ_6	δ_7	δ_8
Predictions	?	+	-	-	+	-	+	?	?
Coeff.	-.006	-.006	-.340	-.282	.244	-.004	.089	-.012	.016
t	-2.683***	-.044	-1.757*	-1.881*	1.969**	-.285	.434	-3.088***	.119

Notes:

1) In this sensitivity test we run the regression using yearly data of 2003 with negative abnormal earnings.

2) ***, **, and * denote significance at the 1%, 5%, and 10% levels, respectively.

3) N = 194, F-statistic = 3.169, and Adjusted R-square = .082.

The variables are defined as below:

CAR	=market-adjusted abnormal returns accumulated over the two trading days (0, +1), where 0 is the annual earnings announcement date, adjusted for dividends and stock rights
ΔE	=unexpected annual earnings, calculated as earnings of this year minus earnings of last year and then divided by market value of equity
T10	=1 if the firm is audited by a Top 10 auditor, and 0 otherwise
SU	=1 if the firm switches to a larger auditor in the current or the last year, and 0 otherwise
SD	=1 if the firm switches to a smaller auditor in the current or the last year, and 0 otherwise
OPI	=1 if the firm receives an unclean auditor's opinion in the current year, and 0 otherwise
MB	=market-to-book ratio at the end of the previous year, calculated as the market value of the firm's stock divided by its book value
ABBETA	=the absolute value of beta to proxy for firm risk

Table 11-3: Market implications of auditor choice and auditor switch for negative ΔE in 2004 --- Test of H7b, H8 and H9

$$CAR = \delta_0 + \delta_1\Delta E + \delta_2 T10*\Delta E + \delta_3 SU*\Delta E + \delta_4 SD*\Delta E + \delta_5 OPI + \delta_6 OPI*\Delta E + \delta_7 MB*\Delta E +$$
$$\delta_8 ABBETA*\Delta E + \varepsilon \qquad (3)$$

	Intercept	ΔE	T10*ΔE	SU*ΔE	SD*ΔE	OPI	OPI*ΔE	MB*ΔE	ABBETA*ΔE
	δ_0	δ_1	δ_2	δ_3	δ_4	δ_5	δ_6	δ_7	δ_8
Predictions	?	+	-	-	+	-	+	?	?
Coeff.	-.016	.010	-.140	-.190	.115	.002	.106	.000	-.012
t	-4.834***	.130	-2.770***	-2.487**	1.786*	.182	1.666*	.610	-.230

Notes:

1) In this sensitivity test we run the regression using yearly data of 2004 with negative abnormal earnings.

2) ***, **, and * denote significance at the 1%, 5%, and 10% levels, respectively.

3) N = 475, F-statistic = 3.985, and Adjusted R-square = .048.

The variables are defined as below:

CAR	=market-adjusted abnormal returns accumulated over the two trading days (0, +1), where 0 is the annual earnings announcement date, adjusted for dividends and stock rights
ΔE	=unexpected annual earnings, calculated as earnings of this year minus earnings of last year and then divided by market value of equity
T10	=1 if the firm is audited by a Top 10 auditor, and 0 otherwise
SU	=1 if the firm switches to a larger auditor in the current or the last year, and 0 otherwise
SD	=1 if the firm switches to a smaller auditor in the current or the last year, and 0 otherwise
OPI	=1 if the firm receives an unclean auditor's opinion in the current year, and 0 otherwise
MB	=market-to-book ratio at the end of the previous year, calculated as the market value of the firm's stock divided by its book value
ABBETA	=the absolute value of beta to proxy for firm risk

Wissenschaftlicher Buchverlag bietet

kostenfreie

Publikation

von

wissenschaftlichen Arbeiten

Diplomarbeiten, Magisterarbeiten, Master und Bachelor Theses
sowie Dissertationen, Habilitationen und wissenschaftliche Monographien

Sie verfügen über eine wissenschaftliche Abschlußarbeit zu aktuellen oder zeitlosen
Fragestellungen, die hohen inhaltlichen und formalen Ansprüchen genügt,
und haben **Interesse an einer honorarvergüteten Publikation**?

Dann senden Sie bitte erste Informationen über Ihre Arbeit per Email
an info@vdm-verlag.de. Unser Außenlektorat meldet sich umgehend bei Ihnen.

VDM Verlag Dr. Müller Aktiengesellschaft & Co. KG
Dudweiler Landstraße 125a
D - 66123 Saarbrücken

www.vdm-verlag.de

CPSIA information can be obtained at www.ICGtesting.com
Printed in the USA
240221LV00008B/7/P